Offshore Wind Summit 2023

Aberdeen, United Kingdom
3 October 2023

ISBN: 978-1-7138-9563-3

Printed from e-media with permission by:

Curran Associates, Inc.
57 Morehouse Lane
Red Hook, NY 12571

Some format issues inherent in the e-media version may also appear in this print version.

Printed with permission by Curran Associates, Inc. (2024)

For permission requests, please contact The Royal Institution of Naval Architects at the address below.

The Royal Institution of Naval Architects
8-9 Northumberland Street
London, WC2N 5DA
United Kingdom

Phone: 020 7235 4622
Fax: 020 7259 5912

publications@rina.org.uk

Additional copies of this publication are available from:

Curran Associates, Inc.
57 Morehouse Lane
Red Hook, NY 12571 USA
Phone: 845-758-0400
Fax: 845-758-2633
Email: curran@proceedings.com
Web: www.proceedings.com

160 YEARS OF HISTORY, KNOWLEDGE AND EXPERIENCE

THE ROYAL INSTITUTION OF NAVAL ARCHITECTS

Offshore Wind Summit 2023

3 October 2023

Aberdeen, United Kingdom

In Partnership With

ABS

TOGETHER WE WILL SHAPE OUR FUTURE

THE ROYAL INSTITUTION OF NAVAL ARCHITECTS

Offshore Wind Summit 2023

3 October 2023

Aberdeen, United Kingdom

THE ROYAL INSTITUTION OF NAVAL ARCHITECTS
8-9 Northumberland Street
London WC2N 5DA

Telephone: 020 7235 4622
Fax: 020 7259 5912

ISSN No: 2752-6569

CONTENTS

CONSTRUCTION OF FLOATING OFFSHORE WIND TURBINES

A P Crowle, University of Exeter, College of Engineering, Mathematics and Physical Sciences, Renewable Energy Group, Penryn Campus, Treliever Road, TR10 9FE, UK. (supervisor Professor P. R, Thies)

SUMMARY

As the floating offshore wind turbine industry continues to develop and grow, the capabilities of shipyard facilities need to be assessed as to their ability to support the expanding construction and installation requirements. This paper assesses current infrastructure requirements and projected changes to shipyard facilities that may be required to fabricate the floating structures. Separate shipyards are required, with large laydown areas, for substructure construction, fit-out of the turbines, moorings, subsea cables and maintenance. The capabilities of established shipyard facilities are assessed by evaluation of size of substructures, height of wind turbine with regards to the cranes for fitting of blades, distance to offshore site and offshore installation vessel characteristics. It is intended that as much work is carried out at a sheltered location inshore but close to the offshore wind farm. This paper reviews the shipyard requirements and defines the role of shipyards for floating wind.

NOMENCLATURE

A nomenclature used in the paper is as follows

[Symbol]	[Definition] [(Unit)]
AHTS	Anchor Handling Tug Supply
Float-out	Substructure floated from dry-dock
FOWT	Floating offshore wind turbine
GW	Gigawatt
HTV	Heavy transport vessel
Load-out	Horizontal movement of substructure from land onto a heavy transport vessel
M	metre
MW	Megawatt
SPMT	Self-propelled modular transporter
SSCV	Semi submersible crane vessel
T	(metric) tonne
UDL	Uniform distributed load

1. INTRODUCTION

This paper reviews the construction and installation of floating offshore wind turbines. The ability to assemble floating offshore wind turbines (FOWT) structures on or near shore means minimising highly weather dependent operations such as offshore heavy lifts and assembly, thus saving time and costs and reducing safety risks for offshore workers. Maintenance might take place in safer nearshore conditions for barges and semi submersibles types.

There is a commercial requirement to minimise the life time cost of energy, i.e. reduce capital and operating costs for floating offshore wind turbines (FOWT). The method of analysis is to use lessons learnt from traditional shipbuilding, from offshore oil and gas construction and from offshore installation of fixed bottom wind turbines. In addition from the concept and early commercial developments of floating offshore wind turbines there needs to be a better understanding of the requirements for shipyards and port facilities.

The floating offshore wind industry is in the early stages of development and shipyard facilities are required for substructure fabrication, turbine manufacture, turbine construction and maintenance facilities.

The function of shipyards for the construction, installation and maintenance of floating wind is varied:

- Supply base for the geotechnical and met-ocean survey of the offshore site.
- Construction of the substructure, possibly an existing shipyard
- Quayside factory for blade manufacture
- Load-out quay for nacelle and tower
- Laydown area for mooring components
- Laydown area for dynamic array cables
- Fit-out quay for installing turbines

- Support harbour during offshore installation phase
- Future maintenance port

The challenges for floating offshore wind turbine (FOWT) depend on the type of substructure, water depth, prevailing weather conditions, seabed soil and size of the wind turbine. To date the largest turbine on a FOWT is 9.6 MW [17]. Larger powered turbines, up to 16 MW, are being ordered for new bottom fixed wind farms.

The remainder of the paper is structured as follows: Section 2 provides a brief literature review for floating wind construction and installation. Options for reviewing floating wind construction shipyards are discussed in Section 3. The analysis of shipyard requirements is considered in Section 4. Results are shown in Section 5. Discussion is in Section 6 and conclusions are given in Section 7.

2. LITERATURE REVIEW

The shipyard for substructure construction can be a long distance from the offshore location. But the fit out yard needs to be as close to the offshore site as possible, to minimise tow out times which are weather restricted. Shipyards are regarded as the pinch points in the deployment of FOWT [1].

The availability of fit out shipyards, and respective water depth alongside the quay, is a major factor in determining the type of floating wind turbine that is to be used. This is important as the design focus for floating wind platforms is often dependent on the offshore wind farm site conditions. Yet, shipyards are a critical engineering constraint to be able to realise the construction and installation of floating turbines.

Upscaling shipyard infrastructure and investments need to be aligned with the long-term use of FOWT [2]. There are several floating wind designs competing for commercial deployment, which will need different infrastructure requirements. The assembly of floating wind systems, as opposed to bottom-fixed turbines, is mostly based onshore. Therefore, shipyards will need expansion of their land area, quay reinforcement, storage for components, cranes and other retrofits to host mass production of floaters and other turbine components.

The floating wind industry could partly use the infrastructure of the existing shipyards, currently used for bottom-fixed offshore wind and offshore oil and gas platforms. Shipyards, existing or future, need to maximise the whole supply chain efficiency of existing bottom-fixed offshore wind platforms. Space is and will become a bigger issue for shipyards. To overcome this, shipyards will require new strategies and regional collaboration.

For example Port Talbot in South Wales [3] is being promoted as a potential shipyard for construction of substructures and fit out of topsides because it has over 15 m of water at low tide in a sheltered location. Shipyard requirements are discussed in [7] for constructing FOWTs in California.

Additional shipyard facilities are needed to support wind turbine manufacturing, substructure fabrication, fit out, and support for offshore installation, operation and maintenance of wind farms. Demonstration scale projects (approximately less than five devices) have different shipyard infrastructure requirements than full commercial scale projects (30+ units), as the supply chain logistics and costs of scale will be significantly different [8].

3. FLOATING WIND STATUS

3.1 GENERAL

The main parts of a FOWT are shown in Fig. 1. The topsides are the blades, nacelle and tower, which are provided by the turbine manufacturer. The substructure is in part limited by the shipyard used for construction and fit out. The moorings include the mooring lines and the anchor type e.g. drag anchors, suction piles, driven piles or drilled piles.

There are dynamic cables from the substructure to the seabed. The export cable is buried in the seabed. The export cable may go to land-based grid, typically via a substation or to supply electricity to offshore oil and gas platforms.

Different types of FOWT are given in Table I. The minimum water depth for alongside quay construction work is at lowest astronomical tide (LAT), with 1 m under keel clearance plus the level trim draft of the FOWT. Shipyards with a water depth of up to 15 m can accommodate semi-submersible, Barge and Tension Leg Platform (TLP) type platforms, whilst Spar platforms require up to 80 m water depths.

Concrete structures, for the same plan dimensions, have a higher weight that steel substructures and so the concrete structures have a deeper draft. Spars both steel and concrete have very deep drafts and gain their intact stability from adding solid ballast to their base. Barges gain their stability from their width and they have the minimum draft, however they have the largest motions in operation. Semi submersibles gain their stability from the second moment of water plane area and have lower motions than barges, multi turbine FOWT are based on very large semi-submersible hulls.

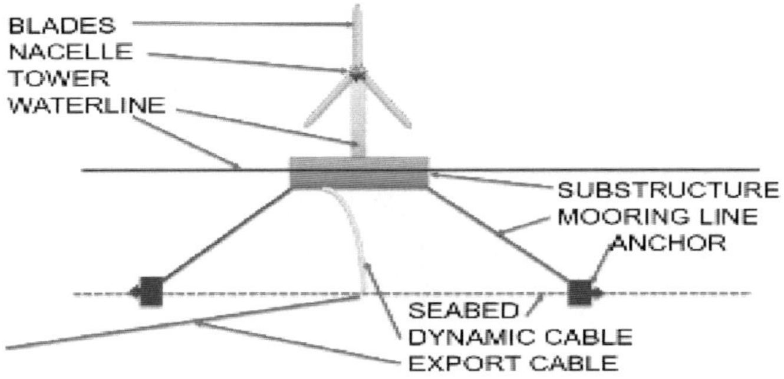

Figure 1: Main parts of FOWT [7]

Table 1. FOWT types fit out quay requirements

FOWT type	Fit out quay Draft range	Substructure construction material
Barge	6 m to 8 m	Steel
Barge	10 m to 12 m	Concrete
Semi-submersible	10 m to 12 m	Steel
Semi-submersible	12 m to 15 m	Concrete
Spar	70 m	Steel
Spar	80 m	Concrete
TLP	10 m to 12 m	Steel
Multi turbine	10 m to 12 m	Steel

The design guidelines for floating wind are provided by classification societies, such as DnV [7].

3.2 BARGES

A barge is a hull made of either steel or concrete, see Figure 2. It is stabilised in place through its buoyancy (water plane area). The assembly of the structure is performed onshore and towed offshore using tugs. Barge structures have a low draft, making them suitable also for shallow water shipyards, though they have higher motions in waves than other types of FOWT. Some barges are anchored to the seabed using catenary mooring lines whilst one type uses weather vanning technology. Currently structures weigh around 2,000 tonnes (steel) and 4,000 tonnes (concrete).

Figure 2: Steel barge [15]

3.3 SEMI-SUBMERSIBLES

A semi-submersible substructure is a hull with columns which are connected to each other with bracings. The platform uses the buoyancy force to provide stability when floating. The structure is anchored to the seabed using catenary or taut mooring lines. This substructure is assembled onshore and, despite its heavy weight, it has a relatively low draft of approximately 10 m prior to installation of the tower, nacelle and blades. The weight of the structure for a single turbine is between 2,500 tonnes to 5,000 tonnes. The most common steel design uses three columns. Some early designs had the turbine at the centre which has the advantage of minimising active ballast systems but has the down side of requiring much larger onshore crane outreach to fit the nacelle and blades. Most current designs have the turbine either in one corner or one side to maximise use, taking account of onshore crane lifting radius.

Other semi-submersible substructure concepts have multi turbines in a single platform. These structures are moored using a weather vanning system.

Figure 3: Semi submersible Load-out from land [12]

Figure 3 shows a load out from land onto a heavy transport vessel. The construction of a semisubmersible in a dry-dock is shown in Figure 4.

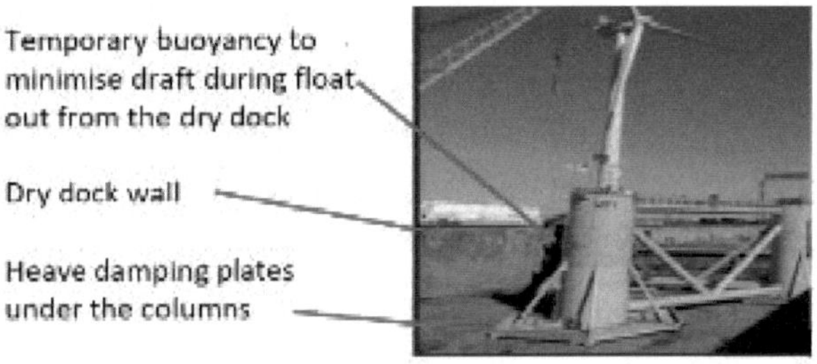

Figure. 4: Semi-submersible in dry-dock [12]

3.4 SPARS

A spar-buoy (or spar) is a cylinder structure. It is stabilised by keeping the centre of gravity below the centre of buoyancy, using a ballast made of one or more heavy materials, topped up with seawter. This is the structure with the largest draft, between 70-90 m once installed, minimising the motions and stabilising the structure. However, this can translate into more complex logistics in the assembly, transportation and installation of the foundation. The structure is anchored to the seabed using catenary or taut mooring lines.

The assembly of the steel substructure is performed onshore by building the spar hull horizontally. The steel weight of one structure is around 2,500 tonnes to 5,000 tonnes before ballasting. The steel substructure is loaded out onto a Heavy Transport Vessel (HTV), taken to deep water in a sheltered location, see Figure 5.

Figure 5: Spar float off before upending [14]

The spar hull is then upended using water ballast. Then solid ballast is added to the base. The turbine is assembled on land and fitted onto the spar hull, in sheltered water using a Semi-Submersible Crane Vessel (SSCV), Figure 6.

Figure 6: Spar topsides lift off quay

The concrete spar starts off by being slip formed vertically in a dry dock. The completion of the substructure continues in deep water using slip forming. Solid ballast and water is added to the base for intact stability. Both steel and concrete types of spar require temporary moorings to be set up in sheltered water and also work barges to be alongside.

3.5 TLPS

A TLP has a high buoyancy force which requires the anchoring mooring lines to be fully tensioned to provide in place stability. However, during the fit out and tow out the TLP has low or negative intact stability and thus unique methods of installation are required.

Options for TLP installation include:

- add temporary buoyancy to the hull
- construct offshore using a crane vessel with active heave compensation
- use a variable draft i.e. large water plane area for float out and a draft with low water plane area during operation

Each of these options will require different shipyard facilities. The advantage of the TLP floater is that it has a lower footprint on the seabed below the structure compared to a catenary moored FOWTs and has lower in place motions. There are three TLPs are under construction, in the South of France, [17]. TLP vertical tension mooring technology leads to complex installation. The weight of this platform can be lower than the semi-submersible types.

3.6 OFFSHORE TURBINES

Turbine blade lengths and nacelle diameters are typical manufacturer's information. Table 2 gives offshore turbine dimensions and weights of the nacelle.

Table 2. Turbine Dimensions and Weights

Power	Blade Length	Hub Height	Total Height	Nacelle Weight
MW	m	m	m	t
8	84	116	202	443
10	94	126	222	579
12	103	135	241	675
14	111	145	260	868
16	118	154	278	1019

The hub height at which the nacelle has to be installed is a limiting factor for available onshore cranes and for existing floating crane vessels. It is assumed that there is an air gap of 30 m between a blade at its lowest point of rotation and the still water surface at low tide.

4. METHODS OF REVIEWING SHIPYARDS

This section gives an overview to the construction and installation phases that need to be accommodated by the shipyard. It is separated into categories as follows:

4.1 SUBSTRUCTURE

The substructure can be built as follows:

- In a dry-dock: This is applicable to barge types, medium sized semi submersibles and the lower part of concrete Spars. Temporary buoyancy may be required to minimise draft and keep trim and heel to zero. There are constraints on dry-dock width and water depth over the sill, and also on availability.
- Or on land: This is possible for barges, semi submersibles, steel Spars and TLPs. A typical yard is a shipbuilding facility or offshore construction site. Load-out is by self-propelled modular transporters (SPMT) onto a submersible barge or HTV. In this case the minimum ground bearing capacity is 10 tonnes/m^2.

FOWT substructure fabrication will generally require buying pre-rolled and weld prepared steel tubulars which require assembly into larger units [12], which in turn needs;

- large storage space
- complex high strength steel welding
- FOWT substructure fabrication is labour-intensive, with many hours of manual welding required.
- cranes for up-righting and gradual assembly.
- large spaces for FOWT manufacture

FOWTs are complicated to fabricate as a serial process because of the large space required for each substructure. Manufacturing facilities may be co-located or spread over a number of highly specialised locations, with additional steps required for ocean transit of goods (i.e., via SPMTs around site and submersibles barges between sites).

There are further practical constraints, such as gaps between orders, meaning that shipyards are not able to retain all the skilled labour hired for a project. Each new job must therefore carry some cost of hiring and allow time for a certain amount of learning on the first units produced, causing process inefficiencies something which is minimised where the same teams are working continuously from one project to the next.

Installation vessels are continuing to increase in size, which is further increasing the available water depth limits for quayside load-outs. Some sites have short windows during high tide to complete load-outs, increasing the load out time and cost.

4.2 BLADE MANUFACTURE

Because of the length of the turbine blades, (see Table II) they need to be manufactured close to a load out quay as they are too long to be transported on public roads. They may be loaded by crane, either individually as shown in Figure 7, or in bundles of three onto a transport vessel.

Figure. 7: Blade load out [16]

Shipyards will continue to increase the use of roll-on/roll-off (RO-RO) vessels particularly for transporting large components (i.e. nacelle, blades). This reduces time and logistical costs compared with traditional methods for component delivery [4].

4.3 LOADOUT QUAY FOR MOORING EQUIPMENT

The mooring system needs to be installed prior to the offshore arrival of the completed FOWT. The mooring lines and anchors will be delivered from their respective construction shipyards to a mobilisation shipyard close to the offshore location. Chain is used for the complete mooring system in shallow water. In deeper water chain is required where the mooring line is on or close to the seabed. Part of the mooring lines may be synthetic fibres or steel wire ropes and would be stored on reels. Specialised installation equipment will need to be stored on land and to be loaded onto installation vessels e.g. work class ROVs, hammers for driven piles, drilling equipment for drilled piles or subsea pumps for suction piles, Figure 8.

Figure 8: Suction pile anchors [14]

The moorings have involve the following marine activities:

- Anchor load out by crane onto cargo ship
- Offload anchor by crane from cargo ship onto mooring storage quay
- Mooring line load out by crane onto cargo ship
- Offload mooring line by crane from cargo ship onto mooring storage quay
- Load onto mooring installation vessel

Mooring and anchor systems can be stored in a separate shipyard and do not need particularly high lifting capability. There is potential to use drums to store synthetic rope, which would require less space.

4.4 LOAD-OUT QUAY FOR SUBSEA CABLES

Subsea cables need to be stored onshore prior to deployment offshore. The cables are usually installed prior to the installation of the FOWT. There are export cables which are buried in the seabed and dynamic array cables which connect the export cables to the FOWT.

The export cables may be built and loaded out a long way from the final location as they require specialising manufacturing facilities. The multiple dynamic array cables can also be built a long way from the offshore location, so a marshalling shipyard may be required for the cables.

A secondary shipyard may be required for concrete protection mat construction and storage. These concrete mats are used to protect the export cable from damage, from dropped objects, trawl boards and ships anchors.

4.5 LOADOUT QUAY FOR NACELLE AND TOWERS

For smaller wind turbines the tower and nacelle may be built away from a loadout quay. However, with the expectation that the turbine minimum size of commercial FOWT is 8 MW, with the likelihood that the wind turbines will be at least 10 MW capacity on FOWT the nacelle and tower will be too big to be transported on public roads.

4.6 FIT OUT QUAY

A fit out shipyard will need access to a large laydown area to store nacelles, blades, towers.

For turbine assembly, a shipyard will need to have cranes capable of lifting the nacelle, the heaviest and highest lifting operation and thus one of the limiting factors. Mobile cranes with sufficient lifting and reach capability are limited in global availability. There are high costs to mobilise as it is transported in sections and needs to be assembled for use. Figs. 9 and 10 show a 9.6 MW being lifted by a large onshore crane. As an alternative, a very large inshore sheer crane might be used for inshore assembly but they are limited in availability, too.

After turbine assembly in shipyard, there is paint touch-up, bolt tensioning checks, electrical circuits and safety system checks to be carried out by technicians. The fit out shipyard requires a large storage area and needs to have a minimum overall strength of 15 t/m^2 as a uniform distributed load (UDL).

Figure 9 shows the lifting of a 675 t nacelle for a 12 MW turbine, [5] wind turbine and blades onto a FOWT [13].

Figure 9: Nacelle lifting [5]

Areas assigned for heavy lifting crane operations must accommodate a minimum surcharge load of 30-40 t/m^2 (UDL) which increases to a maximum surcharge of 50-80 t/m^2 by operation of the main crawler crane on a single track to the load spreading surface, figure 10.

Figure 10: Blade lifting [13]

4.7 WET STORAGE

Substructures need to be wet stored prior to fitting the turbine topsides. This requires the laying of temporary moorings. Similarly, after fitting the turbines more wet storage may be required prior to tow offshore, whilst waiting on weather or available installation vessels. In this case, the temporary moorings have to be stronger than prior to fitting the turbine because of the high wind loads on the topsides.

4.8 NAVIGATION OUT OF FIT OUT YARD

Navigation and fit out shipyard requirements are summarised in Table 3. Regarding the air draft limit there should be no overhead power cables or bridges between the fit out shipyard and the open sea.

Weather forecasts are suitably accurate for up to 3 days in advance. The fit out yard is to be a maximum of 2.5 days sailing time from the offshore location. This allows 0.5 day to get at least 2 mooring lines connected to the FOWT. If the fit out shipyard is too far from the offshore location then a staging shipyard may need to be designated.

Table 3. Fit out shipyard minimum requirements, [6]

Primary Criteria	Semi-Submersible	Barge	Spar
Material	steel	steel	steel
Air draft	No limit	No limit	No limit
FOWT width	90 m	30 m	20 m
Channel width, with tugs	140 m	110 m	90 m
Quay length	120 m	60 m	80 m *
Water depth, at low tide	12-14 m	10-12m	90m
Substructure Shipyard	6 hectare	4 hectare	5 hectare
Area fit out	6 hectare	6 hectare	6 hectare
Fit out quay UDL	15 t/m^2	15 t/m^2	15 t/m^2

*The Spar quay length is for a crane vessel to come alongside the turbine construction site. In addition, the Spar needs a water depth of at least 90m at the sheltered inshore location.

5. RESULTS

Applying the review criteria for each phase a range of constraints and practical requirements regarding weather restrictions, and required shipyard areas have been established.

5.1 WEATHER RESTRICTIONS

The weather restrictions are expected to be as follows for the shipyard activities, Table 4.

Table 4. Shipyard Weather restrictions

Location	Hs [m]	Tp [sec]	Wind [m/s]	Current [m/s]	Visibility [m]
Load-out onto HTV	<1.0	<8	<10	<0.5	400
Float-out from Dry-dock	<0.5	<7	<10	<0.5	400
Float-off HTV	<0.5	<7	<10	<0.5	500
Fit out crane	<0.5	<7	<10	<0.5	300
Wet storage mooring	3.0	10	30	1	500

5.2 SHIPYARD AREAS

The approximate shipyard area requirements for the different factors are shown in Table 5. Shipyard area directly depends on:

- Type of FOWT
- Size of turbine
- Number of turbines in the wind farm
- The number of turbines to be installed in one summer season

Table 5. Shipyard laydown area estimates

Wind farm size	25 MW pre commercial	300 MW commercial
Turbine size	5 MW	10 MW
Number of units	5	30
Tower manufacture	2 hectares	25 hectares
Nacelle manufacture	3 hectares	15 hectares
Blade manufacture	4 hectares	32 hectares
Dynamic cable storage	2 hectares	12 hectares
Export cable storage 50 km length	10 hectare	12 hectares
Substructure construction	10 hectares	100 hectares

It can be expected that as larger turbines become available for bottom fixed structures they will also be deployed on FOWT.

5.3 SHIPYARDS FOR STEEL AND CONCRETE SUBSTRUCTURE

Steel substructures have the following advantages in shipyard construction [9]:

- Transferrable experience from other industries
- Assembly can be executed relatively fast if components are pre-fabricated
- Lighter substructures are possible (compared with concrete) which minimises water depth requirements at the shipyard, dry dock and the fit out shipyard

Concrete substructures have the following advantages in construction shipyards [10]:

- Concrete supply adaptable to local conditions
- There are opportunities for local content
- No specialized equipment,
- Low costs of concrete as a raw material

Steel substructures, [10], have a long history of being used offshore and steel is much easier to recycle. However, steel is subject to corrosion (as is concrete via internal reinforcements, but to a lesser extent). Floating wind substructures made of steel can be assembled faster as these are not as sensitive to environmental conditions (e.g. frost and heavy rain) and are not exposed to concrete curing time

5.4 VISUAL IMPACT DURING CONSTRUCTION

One of the benefits of floating wind is that they will be deployed in locations where they will not be visible from the shore. However there is high visual impact from floating wind turbine during manufacturing, assembly, repair (near shore or in shipyard) and decommissioning. However, it is expected to encounter low opposition as these are short-period tasks that will bring jobs to the area close to the FOWT shipyard, [11].

6. DISCUSSION

The market for floating platforms is still in its infancy. There are many platform types, and only few have been demonstrated pre-commercially. Existing shipyards and fit out quays have been used. Different manufacturing techniques will require different facilities, from large yards to covered premises, to dry docks [11]. Floating offshore structures are large and will need large storage areas and/or transfer to assembly facilities. It is likely that wet storage will be used, so shipyards able to offer sheltered moorings ahead of turbine integration will have an advantage. Storage and assembly may happen at different shipyard locations and could be done at quayside (with onshore cranes) or in sheltered locations using floating assembly bases or floating cranes.

Industrialization of floating wind technology is the key for future cost reductions. Shipyards form an important element in the commercialisation of floating wind and should actively plan for these requirements.

Shipyard facility requirement criteria will differ for each shipyard classification and substructure technology as the functions and installation vessel requirements are different. Because the industry is at an early stage, and deployment technologies and methodologies are still in development, the requirements presented are intended only as a broad review of likely shipyard facility requirements based on available data and technology.

Key outcomes from the research include the following for semi-submersible FOWT construction. Steel barge requirements are similar. Spars are similar but need deep sheltered water close to the shore. TLPs requirements are under development:

- Quayside capacity up to 20 tonnes/m^2.
- A protected harbour
- Floating wind substructures will require deep draft channels.
- For tow-out of an assembled unit there must be no air draft restriction, from bridges or power cables
- Rail and road connections will likely be required for transport of materials

The need for port capacity is increasing, however shipyard development is a long process. In addition, there is a lack of capacity in infrastructure, including cargo vessels, heavy transport vessels, wide roads and trains. The supply chain struggles keep up with the fast developments and everything getting larger, heavier and needing to done more quickly. Cost efficient floating wind turbine substructures are needed for industrial-scale projects. The life time cost of energy is minimized by the low weight of the hull and shallow draft at along the quayside enabling the use of local port infrastructure. Future FOWT need to be based on a modular approach allowing flexibility in fabrication and maximizing the use of the existing local supply chain, whilst reducing quayside and offshore operations to allow units to be installed and fully commissioned at a rapid rate.

7. CONCLUSIONS

Shipyards are important to facilitate the assembly, installation and operations of FOWT. For commercial projects with multiple turbines, shipyards need to be able to construct several FOWT at one time.

For substructure dry ocean transport there are no limits on shipyard locations. However the shipyard requires a quay strong enough for SPMTS to have a side load out onto a HTV. There must also be sufficient water depth for the HTV to stay afloat at low tide with 1m of under keel clearance.

Table 6 summarises shipyard sailing times to the offshore location and functions showing that there up to 17 shipyard/port facilities required.

Table 6. Shipyard Location

	Function of shipyard	Sailing time to
1	Support base for seabed surveys	3 days to wind farm
2	Substructure component fabrication	no limits for cargo ship
3a	Substructure assembly and load out	no limits for HTV transport
3b	Alternate assemble in a shipyard dry dock short distance to fit out quay	
4	Blade construction on load out quay	no limits for cargo ship
5	Nacelle close to load out quay	no limits for cargo ship
6	Tower on load out quay	no limits for cargo ship
7	Wet storage of substructure	1 day to fit out quay
8	Fit out of topside onto substructure	3 days at about 3 knots
9	Wet storage of completed FOWT	1 day from fit out quay
10	Anchor load out shipyard	no limits for cargo ship
11	Chain load out shipyard	no limits for cargo ship
12	Mooring assembly shipyard	5 days to wind farm
13	Export cable load out facility	no limits for cable lay ship
14	Export storage of concrete protection of export cable	5 days for cable lay ship
15	Dynamic array cable load out facility	no limits for cable lay ship
16	Support base for operations including support vessels	2 days to wind farm
17	Support base for offshore installation	2 days to wind farm

The mooring assembly area for anchor systems requires a large laydown area, with adequate mobile crane, before deployment to the offshore wind farm. Mooring lines and anchors require a large space with nearby access to water but do not need particularly high lifting capability. There is potential to use drums to store synthetic rope, which would require less space.

A shipyard looking to take on the final assembly and staging of floating wind projects will need access to a large laydown area to store nacelles, blades, towers. Large space for very big onshore cranes to be assembled and then for there use in lifting FOWT topside components.

For turbine assembly, a shipyard will need a lifting capacity of 1,000 tonnes for the nacelle, the heaviest lifting operation. Mobile cranes with sufficient lifting and reach capability are limited in global availability and to mobilise it is transported in sections and needs to be assembled for use.

Different types of substructures have different shipyard requirements. Shipyard capability is likely to influence substructure design choices. Semi submersibles and barges require large quayside areas (up to 80m x 80m). Spars require a deep-water sheltered area for turbine mating TLPs have a low water plane area and will probably have low stability during towing, so final assembly may take place offshore instead.

8. ACKNOWLEDGMENTS

Alan Crowle thanks his colleagues at the University of Exeter for their assistance in preparing this article.

9. REFERENCES

1. Panel discussion at the Floating Offshore Wind Conference 15-16th September 2021, Aberdeen

2. Report 'Shipyards a key enabler for floating offshore wind sector', Wind Europe, September 2020

3. Associated British Shipyards, Future development potential for Port Talbot

4. A 2030 Vision for European Offshore Wind Shipyards, Wind Europe, 2021

5. 'www.ge.com, 12MW test nacelle, 2019

6. Infrastructure to support offshore floating wind (focus on California), BOEM, 2016

7. 'www.dnv.com, accessed September 2021

8. Determining the Infrastructure Needs to Support Offshore Floating Wind on the Pacific West Coast and Hawaii, BOEM, 2016

9. MATHA, D., BRONS-ILLIGA, C., MITZLAFFA, A., SCHEFFLER, R., 'Fabrication and installation constraints for floating wind', *Science Direct*, 2017

10. PROSKOVICS, R., 'Floating Offshore Wind: A Situational Analysis', *Ore Catapult*, 2018

11. Scottish offshore wind strategic investment assessment, SOWEC, 2021

12. UK Strategic Capability Assessment Offshore Wind Foundations, Offshore wind growth partnership, 2020

13. 'www.principlepower.com,'

14. Benefits Of Floating Offshore Wind To Wales And The South West, Ore Catapult, 2018

15. 'www.equnor.com

16. 'www.bwideol.com

17. 'www.vestas.com,

10. AUTHORS BIOGRAPHY

Alan Crowle has some 53 years experience in marine industries and has BSc in naval architecture and MSc in engineering for marine professionals. His design and construction experience includes offshore fixed and floating structures, surface warships, submarines, submersibles and inshore LNG jetties. He is currently carrying out research at the University of Exeter into the fabrication and installation of floating offshore wind turbines.

COST-EFFICIENT FLOATING WIND DESIGNS (INCLUDING STRUCTURES, MOORINGS AND ANCHORING)

M Goalen, Houlder LTD, UK

SUMMARY

Cost-efficient floating wind is a conundrum – a complicated problem for which no simple solution exists. We must solve it to have renewable energy at the scale needed to survive.

Finding the answer isn't helped by the players waging something akin to a Cold War. An atmosphere of suspicion, secrecy, mistrust and self-interest overriding a shared objective. On the one hand world peace and in the case of offshore floating wind – commercially viable clean energy.

This article introduces the protagonists and their perspectives, explores what we mean by cost-efficiency, shining a light on its hidden complexities by focusing on one crucial aspect of all alternatives (Anchoring and Mooring) and pondering if cost-efficiency is even worth considering right now. It then goes on to suggest how the industry might move forward with some ideas on how we come in from the cold.

1. STATE OF THE INDUSTRY

1.1 THE PROTAGONISTS AND THEIR PERSPECTIVES

Demonstrator projects show that offshore floating wind is technically proven as a feasible concept, the challenge is how can it be delivered for utility scale projects within the budgets the developers have committed to.

For simplicity, let's divide the industry into broad factions: Developers, WTG (wind turbine generator) designers, Floating foundation designers, and the remaining Supply Chain. They share the decarbonising goal but are otherwise in conflict.

The WTG designers are in fierce competition, as such they are very guarded of their designs and innovations. But each WTG has its own limitations. A complicated control system to maximize power generation at different wind speeds, optimal heel and trim angles, allowable accelerations at the Nacelle. They are not keen to share this as it will be a differentiator in the selection process the developer has to make when selecting their WTG. The WTG companies have a revenue stream from fixed wind, but the continuous and unceasing demand for larger WTGs means the R+D budget to design the next turbine to stay in the race is larger than the profits being generated from current sales.

Most floating foundation designers are designing one possible solution to the floating wind challenge. Their budgets are limited, their investment is at risk, and they don't have all the information they require to design their final commercial offering. The design is therefore generic and not tuned for a specific site and WTG combination. At the same time WTG components are still increasing in size and weight and there are unknowns with regards to what a viable mooring solution will be, both factors that will influence the detailed design of any floater. In that respect it is not possible to say floater designs are optimized to maximize cost efficiency. They are yet to complete the detailed design phase.

The supply chain including ports, has a golden opportunity - an industry spanning into the next century that is sustainable and will give people the opportunity to do their bit to heal the world from the impact of global warming. However, this opportunity seems just over the horizon and continuously out of reach. There is not revenue generation available yet for many within that space. Everything is subsidised by QUANGOS or the supply chain themselves in a race to be part of the industry when it starts in anger, but people are getting tired of waiting and starting to look to other areas to keep the lights on between now and when the industry is ready to move. Some are even beginning to question whether the practicalities and costs of offshore wind will make it an unviable venture.

Developers have commitments. They need economy of scale, factory line production, optimized cost efficient fit for purpose designs with a strong supply chain and the skills required to deliver in their timeframe. They want convergence and standardisation across the industry to simplify the problem. They want to buy in fit for purpose 3rd party OEM equipment with warranties to reduce their financial risk of underperforming technology.

So there appears to be a large gulf between where the developers need to be and where the industry is. It will take years of learning, iterating and evolution to go from demonstrator status to refined optimum solutions on a commercial scale. It also makes you question whether the approach of making the design the supply chain's challenge creates a risk of defeating the developers' own purpose by not having mature designs ready to fulfil their needs in time.

1.2 TYPES OF FLOATING FOUNDATION

The complexity is compounded by the variety of floating foundation designs on offer. They can broadly be categorised in four main structure types: spar buoy, tension leg platform, semi-submersible platform, and barge.

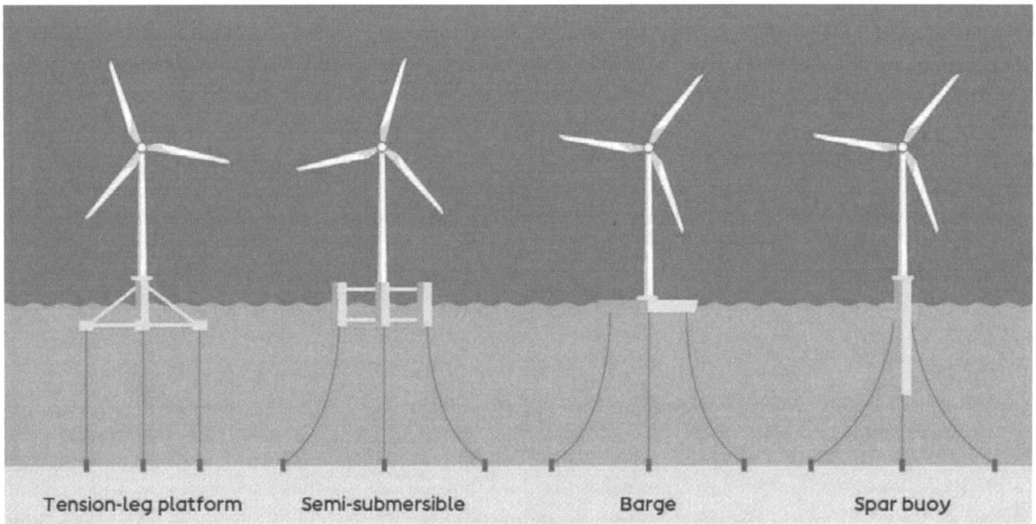

Figure 1: Image courtesy of Orsted

- Spar Buoys - These are very stable given their deep draft with ballast creating a low centre of gravity. They require a deep water area for fabrication and maintenance (if towed to shore). They are conventionally catenary moored which means the mooring lines hang in the water and become horizontal on the seabed. Example Spar - Equinor's Hywind

- Tension Leg Platform - These are less stable than other options until moored. Their biggest advantage is there is less material in the structure so they are lighter and that the taught mooring lines significantly reduce the length of mooring lines needed for deep water locations in comparison to catenary moored structures. Recent developments suggest that TLP do have potential if some of the transportation and installation challenges are overcome. Example TLP - SBM Float for wind

- Semi-submersible platform - These are typically three-legged floating structures. They are probably considered the front runner for most locations given that relatively shallow water is required at the fabrication site, and they are stable for tow during installation, operation and maintenance. Their biggest disadvantage is that the heave and pitch motion is difficult to restrain and can cause large peaks in mooring line loads. Example Semi-sub, Principle Power, Ocergy, and Stiesdal Tetrasub.

- Barge - Square barge structures are being developed which contain a damping pool to reduce and tune the foundation motion. It's a clever concept, and as they can be made from concrete or steel, they could be advantageous for some fabrication locations. Example Barge – Ideol

There are close to 100 different floating foundation concepts in various stages of design. The table below shows the split between each category. It also indicates which are steel and which are concrete, and an average weight per floater.

Table 1. Floating structure design by type

Floater type	%	Steel		Concrete	
		QTY	Average weight	QTY	Average weight
Semi Submersible	61%	84%	4500Te	16%	6000Te
Barge	13%	27%	7500Te	73%	8000Te
Spar	12%	50%	14500Te	50%	10000Te
TLP	10%	88%	2000Te	13%	n/l
Other	4%	100%	n/a	0%	n/a

The table compiled from Quest Floating Wind Energy – Floating Wind Designs of the World [1], shows there are more semisubmersible designs under development than other floating foundation types. This is because of its versatility (maybe the first step of evolution). The semi sub has relatively shallow draft which is good for port access, and temporary storage locations. It also has a clear weight advantage (Te/MW) over other designs apart from the TLP. The semi is also the most adaptable for tuning for a particular motion response. The TLP has more challenges and costs associated with towing, installation and mooring which would need to be overcome before it would be preferrable. In deep water locations the TLP might become more viable.

2. DEFINING COST EFFICIENCY

Cost-efficiency is the ratio of Output to 'Costs Incurred'......but what are the costs incurred?

BV associates recently published document BVGA-16444-Floating-Guide-rl [2]. It illustrates typical costs for a 450MW development (30 x 15MW turbines) over a 25 year period.

The figures contain estimations and averages, there are a few hard to quantify variables excluded but they are good to give an order of magnitude. The table below summarises the total cost over the 25 year period:

Table 2. Floating wind development costs

Section	Name	Cost (£million)	% of Total
P	Project Development and Management	£66	3%
T	Wind Turbine	£600	27%
B	Balance of Plant	£781	35%
I	Installation	£168	7%
O	Operation	£567	25%
D	Decommissioning	£66	3%
	Total	£2,248	

The cost of the floating foundation is rolled up within Balance of plant. An interesting observation is if you were able to optimize the design of a floating foundation to reduce the unit cost by 15% which would be a large optimisation from a design stand point. That would equate to a total saving of £54M or 2.4% saving of total cost.

£54M also equates to approximately 10% of the Operating budget. Design factors that influence the operational efficiency of the floating foundation design are therefore equally or perhaps more important as over a 25-year period with vessels that cost 50-100K per day, and approximately 1.5M to disconnect and reconnect a foundation that needs to be towed to shore for repair it can be reasonably foreseen that equivalent additional costs could be incurred.

3. THE CONUNDRUM

Finding a cost-efficient practical solution that best meets the demands of the developers.

An interesting observation of semi-submersible design is the number of possible pontoon and bracing arrangements that could be viable. Depending on what the designer primary focus was at the scope of requirements phase. So there is a range of traditional round columns, connected with diagonal and horizontal bracings to square columns and pontoons with no bracing, and everything in between. Working in isolation designers have raced to get their design complete and into demonstration phase to increase the TRL level so that they are an option for commercial projects.

However now with the benefit of hindsight academics can compare all possible solutions and determine from analysis the most cost efficient from a cost and stress point of view. (second step of evolution)

Figure 2 below quickly illustrates twelve different possible pontoon and bracing arrangements.

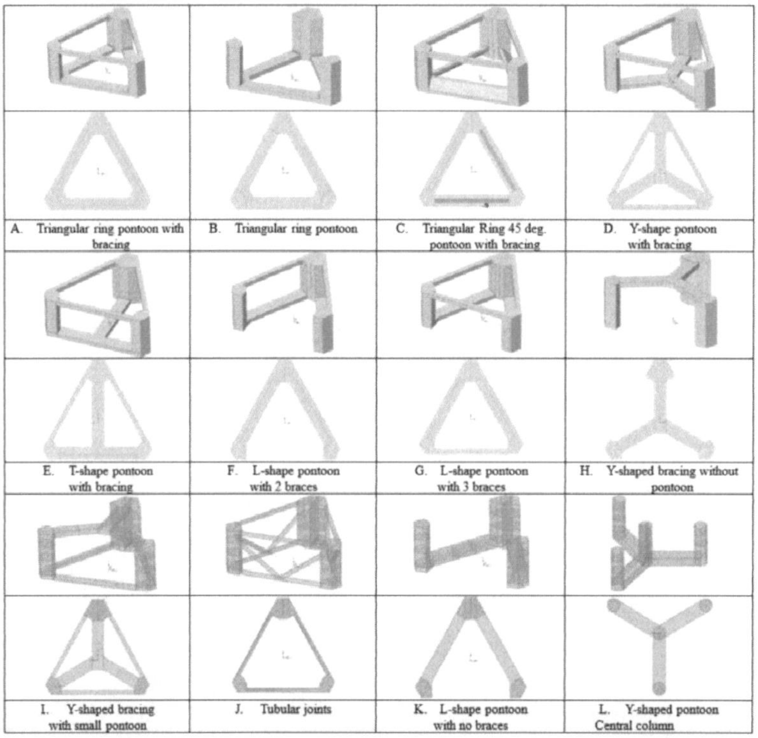

Figure 2: Design Considerations on Semi-Submersible Columns, Bracings and Pontoons for Floating Wind by the Department of Engineering Science and Ocean Engineering, National Taiwan University, Taipei 106319, Taiwan [3].

On top of the theoretical cost efficiency there is a practical side to consider. Access and egress to the floater, connection points for moorings and cables, internal and external inspection of fatigue sensitive connections, and the position of the WTG onboard are all important considerations that will impact the final design. But perhaps the largest driver just now in terms of finding a solution is acceptable to the industry relates to manufacture. Both in terms of the speed at which the structure can be built, and where it can be built in a location with a focus on local content where the supply chain, infrastructure and required skill set is limited or in its infancy.

Concrete structures are therefore still a consideration as is any design that allows for prefabricated sections to be brought to site and assembled quickly. The industry ambition of turning out a floater every two days from an assembly yard is very ambitious and limits design options even more.

Moving from traditional welded connections to pinned connections can allow for quicker assembly but it will take time to evaluate whether there are any detrimental effects during the O&M phase.

There is a difficult balance to find between theoretical cost-efficiency, versus what is achievable locally to each development.

4. FOCUS ON ANCHORS AND MOORINGS – A UNIVERSAL ASPECT OF THE COST-EFFICIENCY CONUNDRUM

For floating foundations other than a TLP, there are three main types of mooring configuration to choose from; chain catenary, semi-taught, taught.

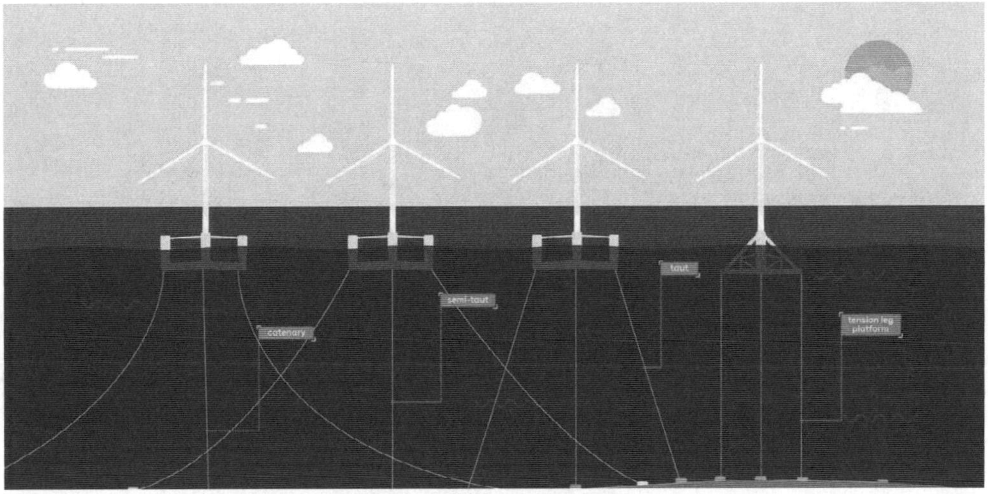

Figure 3: Image courtesy of RWE

The selection of which configuration is most suitable for any development is based on mooring analysis of the chosen floater at a given location to ensure that the SWL of the mooring line is not exceeded for the expected motions, that the floater does not move out with its allowable offset radius and to ensure the motion response of the floating structure keeps the WTG within allowable acceleration limits.

Mooring floating structures with high wind drag in environmentally hostile locations generates high tension in the mooring lines.

Chain is traditionally used in mooring vessels in the marine environment. There are good reasons for this. It has a high resistance to tension but cannot transmit bending forces, and it does not twist or impart torque when pulled tight. The self-weight of the chain provides a restoring force and the weight and length of the ground chain on the seabed means anchors only must withstand horizontal forces.

Mooring analysis conducted indicate chain much larger than current anchor handlers are set up to store or handle will be required. Increasing the number of mooring lines is one obvious answer. That fits with offshore standard practice and increases redundancy in the mooring arrangement.

Taught and semi taught mooring systems are also being considered. These configurations are considered advantageous partly because synthetic ropes are lighter and more easily handled at installation than chain, and partly because there is a requirement for more compliance or stretch in the mooring line to prevent high shock loading when pulled tight. This keeps the motions of the floating foundation within allowable limits for the tower. There does not appear to be a big cost penalty in moving from chain to synthetics either.

These advantages do need to be balanced though. More compliance allows larger offsets and could subject the anchor to vertical loading. Something conventional drag anchors are not designed to withstand.

Anchor selections is driven by, seabed condition, water depth, line load and line angle. The line load is a result of the environmental conditions at a given site while the line angle is driven by the footprint of the mooring pattern and how taught the mooring line is.

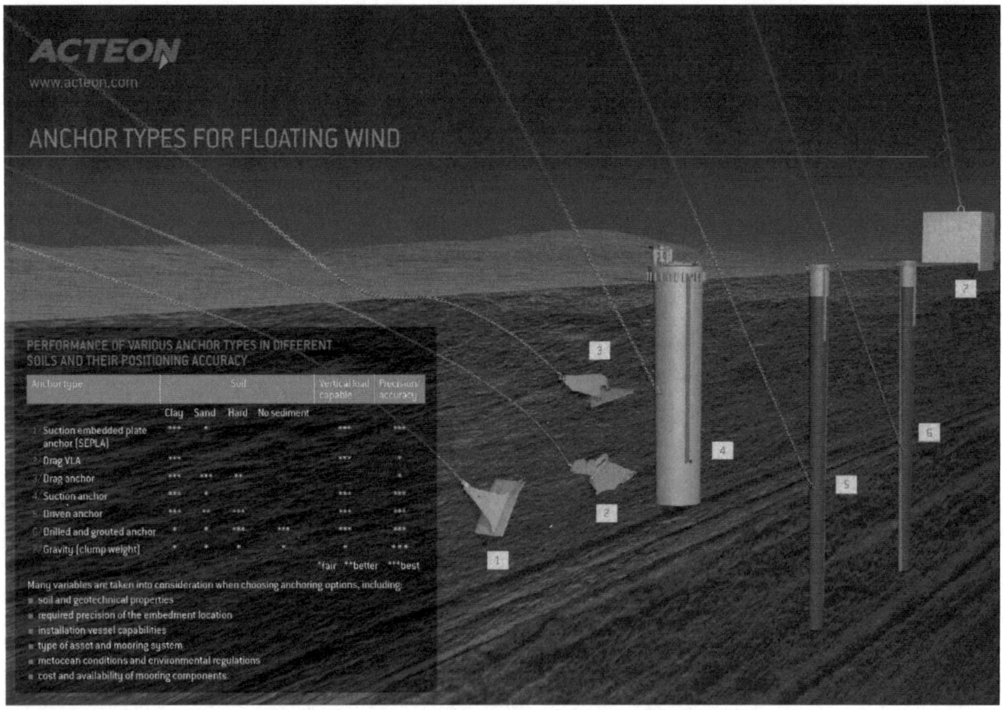

Figure 4: Image courtesy of Acteon Group

The below section compares the different anchor types available on the market.

Table 3. Different anchor types available on the market comparison

	Clump Weights	Large Diameter Drilled & Grouted Piles (LDD&G)	Driven Piles	Suction Piles	SEPLA Anchor	Drag Embedment Anchors (DEA)	Subsea Micropiles
Track Record	Very well understood, and can cope with loads from horizontal to vertical. Troublesome track record.	Very well understood, used extensively in fixed wind. Viewed as a risky option commercially.	Very well understood, used extensively in fixed wind. Require a large vessel spread, risks of refusal. Require high Geotech budget.	Very well understood, used extensively in O&G. Require a large vessel spread, risks of refusal. Require high Geotech budget.	Very well understood, used extensively in O&G. Require a large vessel spread, risks of refusal. Require high Geotech budget.	Very well understood, used extensively in O&G. Requires catenary system and AHV.	Well understood (drilled & grouted pile), novel concept combining with a template. TRL8 in 2023.
Manufacturing Costs	Very high	Very high	High	High, with long lead times	Low	Low	Low
Installation Costs	Very high, prohibitive	Very high, prohibitive	High	High	Medium	High, due to AHV rates and re-installation of 20% of anchors, and chain transportation to site.	Low, 150t crane and PSVs
Load Angles	Horizontal to vertical	Horizontal to vertical	Horizontal to vertical	Horizontal to vertical	Horizontal to vertical	Horizontal only	Any angle
Soil Suitability	Harder seabeds	Broad range of seabed conditions	Broad range of seabed conditions, not suitable for rock or boulder area.	Soft clays, not suitable for rock or boulder area	Soft clays, not suitable for rock or boulder area	Soft clays, not suitable for hard, rock or boulder area.	Any soil
Industrial Solution	No	No	No	No, due to limited soil areas	No, due to limited soil areas	No, due to limited soil areas, prohibitive vessels rates and installation issues at scale	Yes, with 150t crane vessel and PSV fleet 3 x anchors/day installation
Local Content	10%	10%	20%	20%	30%	30%	90%
Project CO2 Emissions	High, due to transport of anchors and large installation vessel.	High, due to transport of anchors and large installation vessel.	High, due to transport of anchors and large installation vessel.	High, due to transport of anchors and large installation vessel.	Medium	High, due to transport of anchors and installation vessel.	Low

The anchor options are displayed in the table in order of cost, highest on the left to lowest on the right. DEA are low cost to manufacture they are well understood and used extensively in marine operations. However, there may be a hidden cost to drag embedment anchors. AHVs (Anchor Handler Vessels) day rates are high and expect to increase with demand. The AHV is used both for installation of the anchor and to load test it once in position. It is expected 20% of DEAs will need installed more than once due to slippage during the load tested. So, there is likely more marine operations in reality than accounted for in initial costings of the development. Drag anchors are also limited in that they are only suitable for certain soil types, and they cannot withstand vertical loading which is prohibitive for taught mooring lines.

It is important to note that water depth and seabed soil conditions can vary within one development. So, while standardisation and convergence of systems is sought it is another example of where things can vary preventing a one size fits all approach.

There could be an emerging disruptive technology that can change that though. Subsea Micropiles claim to be 50% lower than Large Drilled & Grouted which makes them the only anchor type suitable for any soil type including rock.

There is no requirement to load test a fully engineered solution like suction piles, driven piles, large drilled and grouted or micropiles, and the intention is to develop a method of deployment from cost effective vessels of opportunity. Rather than traditional AHVs. All eyes will be on the progress of this technology to see how if it can deliver as promised.

Regardless of anchor type, it is possible to preinstall a mooring system a year or more before the floating foundation is hooked up. Taking that operation off the critical path is one step towards factory line cost efficient strategy. In relation to the floating foundation design, it is important to understand how many mooring lines are required and where they connect to the structure. Consideration should also be given to how easily the mooring line can be connected and disconnected especially if tow to shore is the chosen O+M strategy. Most designers are avoiding using winches to pull the mooring lines in and favour and overboard connection. The water depth that the connection is made is important as ROVs are not good closer to the water surface as wave particle motions can wash them around.

5. COMING IN FROM THE COLD

With many variables that directly and indirectly impact each other, the conundrum is hard to solve. The industry needs economy of scale, factory line production, optimized cost-efficient fit for purpose designs moving towards convergence and standardisation. The reality is we are at the very beginning of evolution towards that. But what could be done to accelerate towards the required status.

Developers - Be bold, select a floating foundation that best fits your scoring criteria that also has the required TRL. Select a WTG and get the designers working together to ensure compatibility. If the development site is also known you can derive everything else from there.

Reality may mean there is additional hardware and marine operations required than allowed for. It is also likely manufacture of the foundations will take longer than two days. If innovation cannot resolve that status the costs will have to be absorbed or the project abandoned.

Reform of the CfD (Contracts for Difference) process might also help. It is a neat mechanism designed to protect the consumer and the developer, but if a developer low bids to secure the contract it puts huge pressure on the supply chain to subsidise the developer's error of judgement. This is then compounded by the relentless drive to reduce LCOE (Levelised cost of electricity). Perhaps everyone should be prepared to pay a premium for sustainable energy for the benefit of future generations?

Governments – Could provide the real financial stimulus up front that's needed to break the deadlock but are currently avoiding doing so. They believe it is the developers' responsibility. Provided the projects do go ahead then lessons learned over time will determine what works and where further evolution is required.

The decision will kick start the supply chain ending the chicken and egg scenario that is so heavily talked about in the industry. Providing confidence to invest and much needed revenue streams.

6. CONCLUSION

Cost efficiency is relative. In relation to the design of the floating structure the scope of requirements will drive the end result. If the critical requirement is to economise and reduce material and therefore cost, the design will look very different from one where the main focus was to assemble 5000Te of steel in a two-day period, and different again to a concrete design to allow for minimal infrastructure and a less-skilled workforce.

It is worth remembering that no single, definitive structure type or floating foundation or mooring configuration will be optimal for every site. There are many variables that will impact the decision, including a technology's operability, reliability, practicality, readiness, CAPEX, OPEX and potential longevity to name a few. A developer must choose a foundation that is most suitable for the wind farm they are developing and, given that many developers' portfolios are global, that will vary from site to site.

Independent, objective strategic and technical consultancy can aid developers in commercialising floating wind – ensuring that costs and risks are minimised, timeframes are realistic, and efficiency is maximised.

Time will tell which are the most suitable options for the industry as years of knowledge and learning are gathered allowing designs to evolve to something close to optimum. It would be natural to find a balance between conservative design used for early-stage development, and disruptive yet to be proven technology.

7. ACKNOWLEDGEMENTS

I would like to thank Simon Hill (Houlder), Colbie Yip (Houlder), Mark Prentice (Intermoor) and Mark Williamson (Subseamicropiles) for their input to this paper.

8. REFERENCES

1. Quest Floating Wind Energy – Floating Wind Designs of the World

2. BV associates recently published document BVGA-16444-Floating-Guide-r1

3. Design Considerations on Semi-Submersible Columns, Bracings and Pontoons for Floating Wind by the Department of Engineering Science and Ocean Engineering, National Taiwan University, Taipei 106319, Taiwan.

9. AUTHORS BIOGRAPHY

Mark Goalen holds the current position of COO at Houlder. The Author is responsible for running the business and its operations. Striking a balance between business development and project delivery. The Author's previous experience includes subsea construction, design and fabrication, vessel operations and maintenance (semi sub operator).

EVOLVING REQUIREMENT FOR FLOATING WIND INSTALLATION VESSELS

P B Lewis and **T Laskowicz**, Intelatus Global Partners, UK

SUMMARY

By 2030, over 6 GW of commercial scale floating wind farms are planned to be commissioned. Installed capacity is forecast to reach over 60 GW by 2035.

The base case assumption for floating wind projects is for large anchor handlers and subsea construction vessels to be deployed to pre-install mooring systems, to tow the structures and to hook-up the floating turbines.

Competing demand from oil & gas projects and an aging fleet present a supply-side challenge. Further, technical factors will challenge much of the existing fleet capabilities and create the conditions for new building demand.

To reflect the unique challenges of floating wind, a new generation of vessel designs are emerging. The economics of such a vessel are challenging, with such a specification expected to cost $150-200 million.

NOMENCLATURE

kV	Kilovolts
MW	Megawatt
GW	Gigawatt
T&I	Transport and installation
EPCI	Engineering, procurement, construction and installation
TLP	Tension leg platfrom
FPSO	Floating production, storage and offloading unit
OSV	Offshore support vessel
WTIV	Wind turbine installation vessels
AHC	Active heave compensated
ROV	Remotely operated vehicle
DP	Dynamic positioning
HVDC	High voltage direct current
TSO	Transmission system operator
IMO	The International Maritime Organisation

1. INTRODUCTION

This paper has been prepared using material from the Floating Wind Turbine Installation Vessel Market Forecast (2023) published by Intelatus Global Partners.

The core objective of this study is the examination of market drivers expected to persist until the year 2035, coinciding with the anticipated large-scale installation of commercial floating wind projects.

The increased interest in renewable energy production is also translating into increased interest and installation targets for the use of floating offshore wind technology. Section two of this paper thoroughly reviews these market drivers and their potential impacts on the industry.

The steps required to build a floating wind farm, including demand for installation vessels are reviewed in section three. The construction process of a floating offshore wind farm has seven elements, which will be analysed in sub-sections.

Section four provides with an assessment of technical factors shaping the requirements for floating wind turbine installation vessels. This section provides an overview of market trends that influence the requirements for installation units. The increase in the number of wind farm components is challenging the transportation and installation capacity of the units.

A summary of commercial conditions with respect to new vessel building activity is presented in the section five. Current market conditions for financing opportunities for new investments in floating offshore wind installation vessels were presented. In addition to technical and financial requirements, investors are also considering environmental issues and the limitations of financing options for projects involved in the oil and gas sector.

Based on the data collected and juxtaposed with development forecasts for offshore wind farms and other sectors, including oil and gas, conclusions are drawn about the expected development of floating wind installation vessels. Summary and conclusions are presented in the sixth section.

2. MARKET DRIVERS

Global offshore wind capacity will grow from around 60 GW at the end of 2022 to 230 GW by 2030 and 470 GW by 2035 driven by the energy transition, security and affordability goals of a growing number of countries.

A significant global wind resource is in deep water requiring floating foundations. To increase offshore wind's share of renewable energy production, developers are taking advantage of waters farther from shore and at greater depths. Floating offshore wind makes it possible to use offshore areas for renewable energy where it would be economically unfeasible to build bottom-fixed offshore wind farms. Around 47,500 GW of technical offshore wind resource is in deeper water suited to floating solutions.

Floating wind is an emerging technology. Currently being tested in small scale demonstration and pilot projects, global floating wind commissioned capacity at the end of 2022 was less than 200 MW. The 11 turbine 88 MW Hywind Tampen project in Norway, fully commissioned in 2023, is the world's largest floating wind farm.

Over 6 GW of commercial scale floating wind farms are planned to be commissioned in Europe and the Asia Pacific Region by 2030. 2030-2035 will see a period a high commissioning activity as the USA joins established European and Asia Pacific markets. Floating wind installed capacity is forecast to reach over 60 GW by 2035.

This demand translates to the mooring pre-lay of over 6,000 mooring spreads and the hook-up of around 5,400 turbines. Close to 25,000 anchors and close to 14,000 kilometres of dynamic inter-array cables will be installed between 2023 and 2035. The available market for vessel owners, $28-145 billion in the forecast period, ranges from pure T&I to full floater EPCI scopes of work.

The forecast is influenced by a variety of factors. Auctions have been identified and awarded and, in some cases, the developers have identified COD dates – but these remain sensitive to new permitting procedures and an immature supply chain.

However, subject to addressing permitting issues, local resistance and transmission challenges, there are significant potential demand upsides in the European Union (particularly Greece, Ireland, Portugal, Spain and the UK), Asia-Pacific (mainly Australia, India, Japan and the Philippines), South America (starting with Brazil) and Canada.

3. BUILDING A FLOATING OFFSHORE WINDFARM

In order to estimate the market demand for the various components used to build an offshore wind farm, the components of the floating offshore wind construction process were analysed. The construction of a floating wind farm can be dived into seven stages.

3.1 MANUFACTURE AND SUB-ASSEMBLY

There are three main substructure types – the spar, semi-submersible and TLP. There are variants of spars (spar buoy and hanging counterweight spar) and semi-submersible substructures (V-column and barge). Substructures can largely be made from concrete or steel. The structures are physically very large, which raises transportation challenges. Till now, only full-scale semis-submersible and spar concepts have been deployed. A full-scale scale TLP pilot array will be commissioned in France next year. In the short- to medium-term, V-column semi-submersible substructures are forecast to be the prevalent solution.

Substructure components are constructed at the assembly port or offsite at local or overseas shipyards, pre-cast bases and/or manufacturing facilities. Final assembly is on land or in drydock.

In the case of spar buoys, they can be built on land or slip formed and ballasted in the water like Hywind Tampen, the temporary works for which required the deployment of three North Sea Barges where mooring anchor test loads were 190 tonnes, 12-22 tonne anchors and 4,250 meters of 76 millimetres chains were installed by Normand Drott with a bollard pull close to 340 tonnes, in what DOF called an operation similar to a typical FPSO hook-up operation.

Manufacturing of the mooring lines, anchors, subsea cables and turbines is carried out in parallel to the substructure manufacture.

3.2 LAUNCHING

Once completed, the substructures are launched over the quayside from land into the water, generally using a submersible barge, supported by harbour tugs.

3.3 TURBINE ASSEMBLY

In the next step, the turbine tower, nacelle and blades are generally installed by large quayside cranes. For example, the Mammoet PTC 200 DS mobile ring crane deployed on Hywind Tampen is one of the world's largest, with the capability to lift up to 5,000 tonnes.

In Japan, we note that Obayashi is repurposing its new building WTIV to support port installation of floating wind turbines. However, we forecast a tight supply of WTIVs for bottom-fixed work and so land based cranes are likely to dominate.

Due to the relative instability of TLPs before station keeping system connection, options are being reviewed to install turbines at site, which will require weather sensitive floating-to-floating lifts or assembling the turbine onto the substructure on a submersible vessel suitable of transporting and launching the assembled structure. These are costly and complex operations.

3.4 MOORING PRE-LAY

Moorings are pre-laid ahead of the floater arrival.

A variety of anchor and mooring line options are available to floating wind project developers:

- Anchors: Drag embedment anchors, suction anchors, gravity installed anchors, vertical load anchors and drilled or drilled piles are among the options that will be chosen for floating wind projects, based on an evaluation of cost, vessel availability and project and site-specific anchor loads and ground conditions.
- Mooring lines: We anticipate that mooring lines will likely feature a top, middle and bottom sections, whereby each section can feature different materials and specifications, including steel chain with single links of up to 220 millimetres, steel wire and synthetic fibre rope. A recent study by NTNU indicates that installation vessels will need to carry and install mooring lines made up of combinations of high and ultra-high strength chain up to 200+ millimetres, 120+ millimetre steel wire and synthetic ropes with diameters from 150 millimetres to over 250 millimetres.

In the short- to medium-term, drag embedment anchors are expected to be the most widely deployed solution, based on supply chain and vessel maturity. Other anchor types will be used, and their deployment will continue to grow as project activity increases and supply chains mature.

A question for the designer of a floating wind farm station keeping system is what happens when a mooring line fails. Mooring design for the floating turbines does not need to meet the same redundancy criteria as manned oil & gas floating platforms. However, the consequence of a mooring failure needs to be considered.

Using a three-line conventional spread by way of example:

- If one line fails but the other two lines are able to support the floating turbine maintaining station, then there is redundancy. In this case three mooring lines per substructure are suitable.
- In the worst case, when a mooring line fails, a floating structure will move significantly increasing stress, to the point of failure, on the remaining moorings or the structure itself. In this case, there is insufficient redundancy and additional mooring lines will be required, increasing the demand for mooring line installation vessel capacity.

The majority of demonstration and pilot projects installed to date have featured one anchor point per mooring line. Equinor is pioneering the demonstration of mutualized anchors. This means reducing the number of anchors deployed on a project. In the Hywind Tampen project, 11 floating turbines are connected by 33 mooring lines (three per turbine) to 19 anchors or 1.7 anchors per turbine. Suction anchors are particularly suited to withstand the force from the multiple directions of a shared mooring system. Drag embedment anchors are not.

Till now, anchor handlers and subsea construction vessels have been deployed for mooring pre-lay operations.

3.5 TOWING AND HOOK-UP

Once completed, turbine and substructure assemblies are towed to site.

The floating structures are hooked up to the pre-laid mooring lines. Vertical loads will be applied by the anchor handler or subsea vessel to tensioners in the mooring lines to achieve the required tension in the mooring system.

The floaters are also connected to dynamic array cables at this time. The array cables can be either pre-laid or installed after the floating turbine is hooked up to its moorings. We are noting a preference from project developers in subsea and flexible cable lay vessels with vertical lay towers to install dynamic array cables. This demand potentially impacts the available supply of suitable subsea vessels for mooring pre-lay work.

3.6 SUBSTATIONS AND EXPORT CABLES

Electricity produced by turbines is transmitted through dynamic array cables generally to a substation for onward transmission to shore.

There has only been one floating wind substation installed offshore, a 66 kV TLP that has since been decommissioned.

For commercial projects, large HVDC substations will most likely sit on moored floating structures familiar to those in the deepwater oil and gas sector or piled jackets located in shallower water than the floating turbines.

As with bottom fixed wind, some markets will adopt radial transmission, where the developer is responsible for building the transmission infrastructure connecting a single project to a shore interconnection, and others, like France will have a TSO that is responsible for offshore transmission, as we have with the EFGL project currently under development by Ocean Winds in France.

4. TECHNICAL DRIVERS

We study ten factors impact the design, construction and operation of floating wind turbine installation vessels in this section. The information presented in the following chapter will then be used in the fifth chapter and will allow a commercial analysis, in particular with regard to the availability of installation vessels that meet certain technical criteria.

4.1 COMPARING A FLOATING WIND PROJECT TO A "TYPICAL" FLOATING PRODUCTION SYSTEM

In the short- to mid-term, floating wind turbine installation is expected to be supported by large anchor handlers and subsea construction vessels as these are largely considered to be the only vessels suited to meet the technical requirements for mooring system pre-lay, towage and hook-up.

Due to the existing supply chain maturity and the fleet profile, floating wind project mooring systems have relied on components familiar to those in the oil & gas industry – anchors, mooring chain, steel wire and synthetic fibre rope.

As such, floating wind farms are currently competing for the same large anchor handlers and subsea vessels as deployed for the mooring pre-lay, towing and hook-up of floating production systems in the oil & gas sector.

Table 1 compares a typical FPSO with a floating wind farm and identifies that floating wind farms will drive demand for a larger number of mooring system installation and towing vessels. In short, the table says that, in this example, there will be 6 to 12 times the number of anchors to install and five times the number of mooring lines in a floating wind project.

Table 1. Comparing a "typical" floating wind and floating production project (Source: Intelatus Global Partners interpretation of information from Equinor, Shell and Delmar-Vryhof)

EXAMPLE PROJECT COMPARISON	1 GW FLOATING WIND FARM	DEEPWATER FPSO
UNITS	66	1
MOORING LINE/UNIT (M)	1,000	2,500
LINE/FLOATER	≥3	16
MOORING LINES (M)	198,000	40,000
ANCHORS	198 (or ~90 shared anchors)	16

4.2 WEIGHT OF THE MOORING COMPONENTS

In the short- to mid-term, the weight of mooring components drives the demand for high bollard pull anchor handlers and subsea vessels with larger AHC cranes.

On floating wind projects, the weight of one drag embedment anchor is expected to be in the range of 20 to 60 tonnes. Drilled and driven piles are heavier, weighing from around 70 to 130 tonnes and suction anchors will weigh in the range of 120 to 180 tonnes, depending on specific project load and soil conditions. The weights of the anchors drive the need for higher bollard pull vessels (250 tonnes and over) and subsea cranes with a sufficient capacity to lift anchor loads at a certain radius and in certain weather conditions to the specified depths of the projects, driving the demand for AHC cranes of 250 tonnes and above.

We have identified that steel chain will be frequently deployed to connect anchors and floating turbines on floating wind projects.

Large anchor handlers have generally been designed to support floating drilling rigs and floating production systems in the oil & gas sector. These vessels generally carry 70-90 millimetres chain for semi-sub drilling rigs and 120 up to in rare cases 180mm millimetre chain for FPSOs. Floating wind projects are anticipated to deploy chain with sizes up to 220 millimetres.

As the size of the chain increases, so does its weight per meter. A 100-millimetre stud link chain weighs around 370 kilogrammes per meter and a 177-millimetre stud link chains weighs around 686 kilogrammes per meter. 220-millimetre chains will have a mass of around 700 kilogrammes per meter.

The heavier chain drives the need for larger chain handling equipment to handle these increased weights during installation operation. This in turn drives the demand for large anchor handlers.

Many large anchor handlers are quipped for a maximum 166-millimetre chain size in the shark jaws, chain gypsies, chain haulers and other handling equipment such as towing pins. Most vessels are accustomed to handling 120 millimetres. Industry sources have indicated that for most large anchor handlers, the effective chain limits are closer to 120-140 millimetres. Above this size, the chain handling equipment on most vessels struggles to efficiently handle the vessel and the result is increased wear and tear on the vessel. Where chain sizes have reached 180-millimetre, modifications to the chain handling equipment have generally been needed.

The Norwegian University of Science and Technology, NTNU, performed a review of vessel requirements to perform mooring system pre-lay for the forthcoming 1.5 GW Utsira Nord floating wind project in Norway, which is currently in the leasing process. Located in an average 270 meters water depth, the scenario that

NTNU based their analysis on features 40 semi-submersible floating turbines, moored with a catenary chain system connected to three drag embedment anchors per turbine. The stud-less chain size in this scenario was 180 millimetres, a link for which has a length of over 1,000 millimetres and a width of close to 605 millimetres. In its study, NTNU concluded that the capabilities of most of the North Sea anchor handling fleet, without modification, the vessel will face significant technical challenges in handling chain of this size.

Of note, one of the advantages of synthetic mooring ropes is their comparative low weight. By way of comparison, a 166-millimetre polyester rope has a dry weight of around 14 kilogrammes per meter and a 211-millimeter rope 33 kilogrammes per meter, significantly lower than that of chain.

4.3 FLEXIBILITY IN HANDLING MULTIPLE ANCHOR AND MOORING LINE DEMANDS

Mooring lines will likely feature a top, middle and bottom sections, whereby each section can feature different materials and specifications, including steel chain, steel wire and synthetic fibre rope.

Large anchor handlers equipped with several winches provide the flexibility to directly deploy anchor chain, spiral strand steel wire and synthetic fibre rope from individual winches over the stern roller of a vessel. Multiple winches reduce the workload on deck to a minimum, enhancing safety and operational efficiency in connecting mooring system sections.

Fiber rope will be preferably spooled from drums/rope reels through a rope tensioner onto the dedicated rope winches ahead of installation.

4.4 PROOF OF LOAD AND MOORING SYSTEM TENSIONING

Till now, demonstration and pilot projects have mostly deployed some of the largest anchor handlers and construction support vessels for mooring pre-lay and hook-up activities, those with a bollard pull or an AHC subsea crane capacity of 250 tonnes or more.

After installation, anchors are proof loaded for a period of time to verify a sufficient holding capacity, which is expected to be as high as 1,000 tonnes on commercial-scale projects.

Post-installation, conventional spreads can be tensioned with inline and seabed tensioners, multiple anchor handlers or cross tensioning against a reaction anchor.

For an inline tensioner, the vertical pulling force applied only needs to be around 40% of the horizontal pretension loads. Requirements for 1,000 tonnes loads drives anchor handlers with bollard pulls of around 350 tonnes and subsea cranes of 400 tonnes. These vessels are deployed to support the oil & gas industry and their utilization is relatively high.

In a study, NTNU assessed the capabilities of the Normand Drott, one of the largest anchors in the world, which features a bollard pull of 339 tonnes. The vessel's special handling winch, designed to accommodate a combination of large loads and greater water depths, has a pulling capacity of 500 tonnes. Using an inline tensioner, the vessel can tension an anchor to around 1,250 tonnes.

4.5 CLEAR DECK SPACE

An important driver for vessel selection is the ability to transport a large number of anchors on deck to the installation site. For this driver, the type of anchor chosen has an impact on vessel carrying capacity. Drag embedment anchors have the smallest footprint and around 12 anchors can be carried on an 800 square meter back deck found on the largest anchor handlers. Piles require around three times the area of a drag embedment anchor and suction anchors four times.

Using Table 1, an average anchor handler of 250 tonnes bollard pull and above will require around 17 trips to transport 198 drag embedment anchors, 40-50 trips for piles and 60 trips for suctions anchors. For the FPSO, two trips are required for the transport of drag embedment anchors, three to four trips for piles and five trips for suction anchors. The number of trips has a significant impact on the cost of installation due to the time required to charter the vessel and hire crews, as well as fuel consumption. With a larger deck area, the number of trips can be minimized.

Only a small number of large anchor handlers have back decks much bigger than 800 square meters.

4.6 CHAIN LOCKER AND CHAIN HANDLING CAPACITY

Using our comparison project from Table 1, we note that that the length of mooring line required is around five times greater in the floating wind project than the FPSO project. Based on average large anchor handler capabilities, this translates to around 12 to 25 journeys depending on the size of chain selected. Up to 13 journeys are estimated for vessels carrying 130-millimetre chain and 24 to 25 for 177-millimetre chain. The range for the floating wind project is estimated at between 50 to 90 trips, which is roughly three-and-a- half to four times more trips.

Large anchor handlers have generally been designed to support floating drilling rigs and floating production systems in the oil & gas sector. These vessels generally carry 70-90 millimetres chain for semi-sub drilling rigs and 120 up to in rare cases 180-millimetre chain for FPSOs.

The majority of chain lockers for the large anchor handler fleet fall in a range of 500 to 1,250 cubic meters at 100% storage utilization. Using 75% utilization factor, the average chain locker of the larger anchor handlers is in the 570-710 cubic meter range.

Figure 1 presents an analysis of the chain locker capacity of the existing large anchor handling fleet for three different chain sizes. Using Hywind Tampen as a "typical" project, we note that each turbine is moored by three 950-meter mooring lines, of which the steel chain element is 730-meters. As such, one spread contains 2,190 meters of chain. Assuming 130-millimeter chain, that translates to close to 2,400 cubic meters of chain volume. Only one active anchor today can accommodate more than two "typical" Tampen spreads at a time (assuming 75% stowage factor).

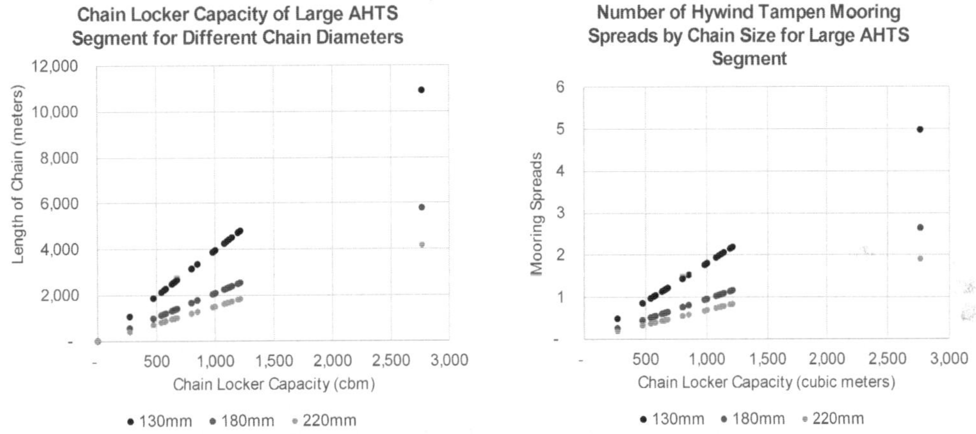

Figure 1: Chain locker analysis (Source: Intelatus Global Partners)

As the chain size increases then the chain lockers accommodate a smaller amount of chain and therefore more journeys are required.

4.7 OTHER MOORING AND PRE-LAY TECHNICAL DRIVERS

Floating wind operations will see the deployment of heavy anchors, chain, tensioners and other jewellery over the stern rollers of anchor handlers. Damage to the stern rollers is a possibility. Several large anchor handlers are equipped with anchor handling A-frames that can be deployed at an angle sufficient to deliver components such as inline chain tensioners over the stern whilst minimizing vessel wear and tear. To ensure sufficient clearance, the anchor handling frames are located at or aft of the towing pins and shark jaws.

Vessels that also support the deployment of work class ROVs will be attractive. The ROVs will support various pre-lay, hook-up and tensioning roles. At least two ROVs are preferable to support operational flexibility and redundancy. The ability to manage deployment during worse weather through a vessel moonpool has additional advantages.

4.8 TOWING AND HOOK-UP SUPPORT

In the demonstration projects and pilot arrays completed to date, turbines have either been mated to the substructures at ports using large land-based cranes or in sheltered anchorages using the largest heavy lift semi-submersible vessels. This methodology remains the base case for most developers.

Alternatives to this base case are being reviewed. There have been several research projects investing the possibility of installing wind turbines and towers offshore at the floating wind farm sites. Solutions include using large heavy lift semi-submersible vessels and DP3 floating turbine (and foundation) installation vessels.

Demonstration and pilot projects have mostly deployed large anchor handlers and towing vessels which also support station keeping during hook-up. Vessels with a bollard pull of at least 200 tonnes is the trend for main towing vessels. The largest vessels deployed have featured bollard pulls above 300 tonnes. At least two smaller anchor handlers (above 150 tonnes bollard pull) have generally been deployed to support towing operations in early projects.

According to DOF, the tow of one 8 MW Hywind Tampen spar-based turbine required 100 tonnes of sustained towing force. When factoring in weather and sea conditions, the result was a requirement for over 200 tonnes of lead tug bollard pull.

Maintaining station of the floater after tow and during hook-up of the floater to the mooring system and tensioning of the mooring system has involved a spread of vessels. These have included some larger AHTS (over 300 tonnes bollard pull), towing and assistance tugs (above 150 tonnes bollard pull) and ROV deployment vessels (either the AHTS or subsea vessels).

4.9 ENERGY EFFICIENCY AND CARBON INTENSITY

As global focus on impacting the harmful impacts on health and the environment from marine industry emissions grows, the regulatory landscape on energy emissions and carbon intensity is evolving and may challenge older vessels.

Three factors are influencing how vessels will be operated going forward. Traditionally biased towards deep-sea and short-sea commercial shipping operations carrying dry and wet cargos, containers and passengers, these factors are increasingly impacting operators of offshore support and construction vessels and underpins interest in more energy efficient vessels and those with lower carbon intensity.:

- Global, regional, national and local regulation on energy efficiency and carbon intensity. These are discussed in more details below.
- Finance and insurance requirements: There are several "*green finance*" initiatives covering debt finance and bond issuing that consider the environmental impact of the activity. Regular carbon intensity reporting, generally based on IMO guidelines has become a feature of lending criteria for a number of traditional shipping banks and insurance companies (including their investment arms). As a result of enhanced sustainability reporting, shareholders are also influencing the investment decisions and operational activities of listed energy companies, suppliers,contractors and their supply chains.
- Customer preferences for low and zero carbon operations.

At a global level, the IMO promotes energy efficiency and carbon intensity in the maritime sector. Till now, these measures have not always impacted the offshore support fleet. However, changes to the IMO strategy on maritime emissions is expected this year and may impact vessels required for floating wind farm installation.

At regional level, the European Union has included offshore vessels in its latest compliance regime. Effective 2025, offshore vessels above 400 gross tonnes will comply with the European Union Monitoring, Reporting and Verification reporting requirements. By end 2024, vessel owners will be reporting carbon dioxide, methane and nitrous oxide emissions. Offshore vessels above 5,000 gross tonnes will be included in the European Emissions Trading Scheme from 2027, which taxes carbon emissions.

Outside of the IMO and the European Union, there are country level requirements to consider, such as those of the UK and Norway and even those at a regional level. An important example of local regulations can be found in California. The California Air Resources Board's Commercial Harbor Craft regulation defines emissions standards for vessels calling into state waters and are stricter that the federal Environmental Protection Agency's regulations. Stricter amendments are currently under review. These will likely impact the anchor handlers involved in California offshore wind projects towards the end of this decade.

Based on the technical assumptions, listed in Section 4 of this paper, there are around 40 anchor handlers that we define as optimum for floating wind projects i.e., they have a bollard pull of 250 tonnes or above and a back deck of 800 square meters and above. The optimal subsea fleet, with AHC cranes of 250 tonnes and above, amounts to close to 100.

In the last seven years only ten large anchor handlers and subsea construction support vessels have been delivered.

By 2030, when floating commercial wind farms are expected to be installed in greater number, only 9 large anchor handlers will be 15 years or younger and 7 subsea construction support vessels.

The existing fleet is almost exclusively powered by internal combustion engines running on conventional fuels. Some vessels feature post-combustion after treatment. Dual-fuel engines and battery energy storage systems are not widespread.

4.10 HUMAN RESOURCES

The last decade has seen the loss of a significant amount of shore-based and mariner competency as a result of low fleet utilization and financial pressures. A further trend has been the increasing influence of local content requirements in key markets.

The operation of large anchor handlers and subsea vessel is a highly specialized activity, calling on experienced teams to ensure projects are completed safely, on-time and within budget.

Demand for qualified and experienced personnel for vessel operations will grow in both the floating wind and oil & gas segments. Some operators have raised concerns about the longer-term availability of competent personnel.

5. COMMERCIAL CONSIDERATIONS

5.1 THE CONDITIONS FOR NEW BUILDING

Conditions exist for potential new building to be considered.

As noted, there has been limited new building activity of large anchor handlers and subsea vessels suitable for floating wind in recent years.

Five 260± tonnes bollard pull anchor handlers, ordered at the end of 2014 were delivered between 2017 and 2019. The five-vessel order was reported at around $80-85 million each. A 477 tonnes bollard pull Island anchor handler which is also equipped with a 250 tonnes AHC crane was delivered in 2020, the estimated CAPEX of which is $170 million. The same vessels today will cost substantially more. Due to the increase in material and labour costs, the order of such a vessel in today's market conditions requires a significantly higher capital investment.

Brokers indicate that new build slots for a subsea construction vessel with a 250 tonnes crane is around $100 million and will rise to around $145 million for a vessel with a 400 tonnes crane.

5.2 NEW GENERATION VESSEL CONCEPTS

Designs are emerging for a new generation of vessel dedicated for floating wind mooring pre-lay and hook-up. Estimates for these new vessels range from $150 million to over $200 million with a delivery time of around three years. Common features of these new generation anchor handlers are:

- Flexibility to install different anchor types.
- High bollard pull.
- Large active heave compensated subsea cranes.
- Multiple large winch drums to provide the volume and flexibility in operation, handling chain, spiral strand wire and synthetic rope.
- Large chain lockers capable of handling large size chain.
- Large back decks of at least to carry drag embedment, suction, piles or other anchor types.
- Work Class ROVs and a moonpool to support anchor deployment and hook-up operations.
- Under deck space for mooring equipment storage, such as buoys.
- Flexibility to support different tensioning options.
- Fitting of an anchor handling frame.
- Low or zero emissions operations coupled with fuel flexibility.
- Embedded digitalization to support, as a minimum, enhanced emissions reporting.

5.3 THE FINANCIAL CASE FOR NEW BUILDS

Offshore support vessel owners and their financial partners continue to deal with the damage of low utilization post 2014. Many owners experienced financial difficulties associated with low oil prices and were unable to pay down debts accumulated during new build expansion programs initiated between 2008 and 2013. Many of the traditional shipping banks, leasing companies and investment funds that funded the new building boom took significant impairments which have left them very cautious about lending to OSV projects. Many private equity vehicles, mostly created in Norway, Germany and the UK, suffered heavy losses from OSV investments.

Finance is reported to be still available from new market entrants and alternative capital providers. However, the appetite for financing new OSVs comes with many restrictions. These include higher pricing of debt based on a lack of historical data for floating offshore wind project support, requirements for long-terms charters to mitigate some bank risk, requirements for corporate recourse and cashflow from charters of at least three to five years depending on the size of the borrowing company. Interest rates, which have risen significantly in almost all global economies, are another factor affecting the cost of capital. Higher equity levels are expected as leverage is not expected to exceed 60%.

Further, as we have noted previously, there is an increasing influence of environmental, social and governance criteria on bank lending. Whereas a bank may be willing to finance an OSV dedicated to floating wind projects, it may face restrictions in lending to a project that also undertakes oil & gas projects. This is a challenge given the structure of the large anchor handling segment. According to brokers, historically there was a general work split of one third spot work / two-thirds term charter in the large anchor handling segment. Today it is closer to 50-50. Further, offshore wind projects are likely to be campaign based, where activity only occurs in the months with good weather. This leaves vessels idle for several months and in need of finding other work, either in the sector or in other regions. This challenges the financial case for vessels designed and built specifically with floating wind projects in mind.

In the North Sea, average term rates for anchor handlers of around 250 tonnes bollard pull and above are only now reaching levels last seen in mid-2024 and are reported by brokers to be around $60,000 per day.

After paying vessel operating expenditures, corporate overheads and debt service obligations, a company is likely to make a significant financial loss on a new building anchor handler. In short, higher day rates and higher utilization than are currently achievable in the market are required to trigger new building.

6. CONCLUSIONS

A relatively immature technology today, commercial floating wind farms will begin installation at scale in the latter part of this decade. At the same time, oil & gas activity is expected to remain strong.

Vessel characteristics familiar to the oil & gas segment will also be required to support floating wind development. Demand for large anchor handlers and subsea construction vessels will grow. Vessel shortages are likely in the longer-term.

There are around 40 large anchor handlers active today that meet the minimum optimal requirement for commercial floating wind farms and a further 100 subsea vessels. However, despite being suitable for floating wind, the majority of the current fleet lack one or more of the technical characteristics required to satisfy all technical drivers.

The market conditions are in place for interest in new building activity. However, commercial challenges remain for both traditional optimal vessels as well as for new generation floating wind installation anchor handlers. When these commercial barriers are addressed, new vessel construction is likely.

7. REFERENCES

1. Intelatus Global Partners, 'Floating Wind Turbine Installation Vessel Market Forecast', 2023.

2. HASLUM, H., (Equinor), 'Developer and perspective and lessons learnt', 2020.

3. LUNDE, T H., ROKSVAAG, T B., and SOLHEIM, S., 'Mooring of Floating Offshore Wind Turbines A review of the mooring designs and the installation vessels for floating offshore wind turbines', June 2021.

4. SJASTAD, H V., (Norwegian University of Science and Technology), 'A simulation study of Installation Concepts for Floating Offshore Wind Farms', 2020.

8. AUTHORS BIOGRAPHY

Philip Lewis is the Research Director for Intelatus Global Partners, a consulting and business intelligence company specialized in the offshore and marine segments. Philip has over 30 years of corporate and commercial management and strategic planning experience in advisory, senior leadership team and Board positions, developing strategy and driving change within the renewable energy, offshore marine, shipping and heavy engineering space.

Tomasz Laskowicz is an intern Research Analyst with Intelatus Global Partners is currently completing his PhD at the University of Gdansk in Poland. Tomasz's PhD in the field of offshore wind focuses on the social acceptance of offshore wind farm development and local content issues within the supply chain.

MARITIME INTEGRITY OF FLOATING OFFSHORE STRUCTURES

R Gopalakrishnakurup, UK HSE, UK

SUMMARY

Operating in harsh environment such as North Sea brings enormous challenges in maintaining maritime integrity of Floating Offshore installations. Some of the key issues include maintaining watertight integrity and stability, integrity of mooring/station keeping systems, bilge, and ballast systems etc. A review of some of the maritime integrity challenges that these installations face will be presented along with some insight into current industry practices and guidance on managing maritime integrity through design, operation, maintenance, and inspection.

NOMENCLATURE

A nomenclature is required for papers/technical notes using many symbols, abbreviations, and acronyms. SI units should be used wherever possible and ordered alphabetically.

ALS	Accidental Limit State
FPSO	Floating Production, Storage and Offloading Unit
FPU	Floating Production Unit
FSU	Floating Storage Unit
FOWT	Floating Offshore Wind Turbine
HSE	Health & Safety Executive, UK
IMO	International Maritime Organization
PSA	Petroleum Safety Authority, Norway
REWS	Radar Early Warning Systems
ULS	Ultimate Limit State

1. INTRODUCTION

The primary function of a floating offshore structure is to provide a stable platform to fulfil its operational objectives. The objectives could vary between a floating, production, storage, and offloading unit (FPSO), floating production unit (FPU), drilling unit, an accommodation unit or even a floating offshore wind turbine (FOWT). Depending on the operational profile of the unit, a floating structure must provide sufficient buoyancy to support its operational objectives, stability against capsizing for the anticipated loading conditions, positioned within an acceptable operational envelope and be able to withstand foreseeable loads acting on the structure during its operation. While some of these structures could be permanently positioned at a location with a pre-defined operating life, others may be located for a fixed period to complete specific tasks.

Most common floating structures are ships, and ship designs continue to evolve to meet our ever-changing operational demands. Design of ships have traditionally been governed by Classification Society Rules based on empirical formulations developed from engineering judgement and operational experience. Classification Society Rules have accumulated more than 150 years of operational experience gathered from the design, construction, and operation of thousands of ships. These Rules are mostly prescriptive in nature and based on accrued knowledge of failure modes. Acceptance criteria provided in Class Rules and other international codes have been periodically amended considering learnings from various accidents. From the Titanic disaster to Deep Water Horizon incident that occurred nearly a century apart, hundreds of incidents have involved shaped current marine design codes.

Designs that extend beyond the boundaries of existing knowledge and experience will require a robust risk assessment to identify unique failure modes, and a rational assessment to check against acceptance criteria based on reliability concepts. Design of offshore floating structures deployed in North Sea is generally subjected to similar process of assessment against site specific environmental conditions. Such assessments are often probabilistic, first principles based, and consequence oriented. The consequences under consideration could be a combination of loss of lives, environmental incidents, financial loss as well as reputational damage.

Quality control during the construction and installation phases are crucial to ensure that the structure is constructed and installed as envisaged, and meets the assumptions made during the design phase. It is essential to ensure that specification of structural components and other equipment meets the intended design specification, fabrication and installation tolerances are within acceptable limits and non-destructive testing of fabricated components confirms maximum dimensions of defects are tolerable.

By virtue of the design configuration, unlike a bottom fixed offshore structure, a floating structure is vulnerable to the way it is operated and require regular human intervention to ensure that they operate as designed. Often the human intervention is in conjunction with control systems which whilst optimising routine performance may complicate the human response to abnormal events. Errors arising from human intervention have led to some significant incidents. This is generally managed through robust engineering controls, operational procedures, and deployment of competent personnel.

During the operational phase of the offshore structure, it is important that the original design intent is maintained. This may involve developing a risk-based assessment taking into consideration the unique characteristics of the offshore structure design, operational environment, and operating profile, and implementing a suitable inspection, maintenance, and repair process. A robust integrity management process is required to maintain the structure within its original design premise.

Overall, integrity of a floating offshore installation is assured only through a combination of design, quality control during construction, assembly and installation, implementation of a robust integrity management system, operational procedures, and competent personnel.

2. DESIGN PRINCIPLES

In UK, the Offshore Installations and Wells (Design and Construction, etc) Regulations 1996 (DCR 1996 [1]) governs the legal requirements for the design of an offshore installation. These regulations require the duty holders to ensure that the designs are such that, so far as is reasonably practicable, it can withstand such forces acting on it as are reasonably foreseeable and its layout and configuration will not prejudice its integrity. It is a legal requirement that fabrication, transportation, construction, commissioning, operation, modification, maintenance, and repair of the installation may proceed without prejudicing its integrity. In the event of a reasonably foreseeable damage to the installation it will retain sufficient integrity to enable action to be taken to safeguard the health and safety of persons on or near it. Only structures that meet the definition of an offshore installation under Regulation 3 of Offshore Installations and Pipeline Works (Management and Administration) Regulations 1995 (MAR 1995 [2]) come under the remit of DCR, and do not include offshore renewable structures. However, principles of the regulation can be applied to other offshore structures including FOWT.

Key elements of the requirements are;

- Ability to withstand reasonable and foreseeable loads, normal and accidental.
- Ability to withstand foreseeable damage and
- Maintaining integrity through various phases of life cycle.

2.1 FORESEEABLE LOADS AND FORESEEABLE DAMAGE

Reasonably foreseeable forces are those forces that a competent person in the field, exercising reasonable foresight, would identify in the light of current understanding. Such loads are expected to include both normal and accidental (abnormal) forces or loads. Broadly, the application of normal loads would not be expected to result in significant damage (or loss of integrity) to the installation. In contrast, accidental loads may in some cases be expected to result in some damage.

In most cases, established industry standards such as ISO 19900 suite of standards or Classification Society Rules can provide a guidance on general requirements and foreseeable loads that need to be considered in the design of offshore installations. Normal loads correspond to the Ultimate Limit State (ULS) and accidental or abnormal loads correspond to Accidental Limit State (ALS) conditions defined in these codes. Typically, normal loads for ULS design checks correspond to 10^{-2} annual probability of failure and ALS checks correspond to 10^{-4} annual probability of failure.

For most offshore structure specialists involved on a day-to-day basis on the design, construction, installation and operational integrity management of floating offshore installations, compliance with Class Rules, International Standards and Codes and other industry guidance may be adequate. Industry guidance, especially Class Rules may look empirical, but have evolved over 150 years of operating experience, addressing any short comings whenever there is any evidence of accidents occurring as a direct result of design shortcomings. Following major incidents involving floating structures, industry has come together to rectify those deficiencies leading to those disasters. Whether it is the oil and gas installations (Ocean Ranger, Alexander Kielland, Deepwater Horizon) or ships (Titanic, Herald of Free Enterprise, or Estonia), accidents have prompted the industry to examine current requirements and make improvements to prevent recurrence. Classification Society Rules thus developed for ships have now evolved to apply limit state design approach for floating offshore structure design and assessment.

Even though for the majority of offshore structures, the design of an installation based on current good engineering practice will be adequate, any novel designs or operating profile should, however, be appropriately risk-based and compliance solely with existing codes, standards and guidance may be insufficient. For a floating offshore structure, the following failure modes must be considered from a maritime integrity perspective, as these can affect safety, environment as well as energy security.

- Loss of stability (capsizing)
- Loss of watertight integrity (sinking)
- Loss of structural integrity and
- Loss of position and/or heading

Unlike a fixed offshore installation, failure of a floating offshore structure can affect safety of people not just on the installation, but other users of the sea including people on other installations and ships in its vicinity. A drifting installation poses risk to other surface and sub-surface infrastructure as well as ships, thus threating safety, environment, and energy security.

2.2 STABILITY & WATERTIGHT INTEGRITY

A stable floating installation is the outcome of a combination of appropriate structural configuration (dimensions and proportions, subdivision, location, and type of openings), underwater hull form and weight distribution supplemented by a ballast system operated by competent personnel that is adequately configured to support changes in loading conditions and a bilge system that can deal with foreseeable water ingress into buoyant compartments. Survivability of a floating installation is achieved by compliance with a set of intact and damage stability criteria that sets out minimum requirements for initial metacentric height, dynamic stability, minimum range of positive stability and angle of down flooding. HSE Research Report RR387 outlines requirements for intact and damage stability for different types of floating offshore installations. Provisions of International Maritime Organization (IMO) conventions such as International Convention for the Safety of Life at Sea (SOLAS) and International Convention on Load Lines and codes such as Code for the Construction and Equipment of Mobile Offshore Drilling Units (MODU Code, [13]) and International Code on Intact Stability (IS Code) provide standard of construction, subdivision, and watertight integrity.

Maintaining watertight integrity through various phases of operating life of a floating installation is achieved through a rigorous program of monitoring and preventive maintenance routines, calibration of equipment, inspection including non-destructive testing as well as functional checks. Equipment and fittings thus requiring inspection, maintenance and testing include ship side valves and associated pipework, ballast system including ballast pumps, valves, piping and control system, bilge system including bilge alarms, watertight and weather tight closing devices and instrumentation such as draught gauges and tank level indicators. HSE Research Report RR387 [6] provides general recommendations for watertight integrity and Offshore Information sheet OIS 4/2011 [9] provides guidance on inspection of ship side valves, flood detection and control as well as inspection and training to maintain watertight integrity.

Bilge and ballast systems play a crucial role in managing stability of a floating installation, especially for a semi-submersible hull form. A review by HSE in 2006, RR473, [6] on the issues associated with stability of semi-submersibles identified that "ballast system failures, errors in operating the ballasting system, and flooding due to variety of causes, including human error, seem to be the most important issues". HSE research report RR387 prescribes considerations for the configuration and capability of ballast system for column stabilised units to function under normal operating circumstances as well as under accidental flooding scenarios.

DCR requires an installation to maintain sufficient integrity in the event of foreseeable damage to the installation. Hull damage and resulting loss of buoyancy or watertight integrity is a credible scenario. HSE Research Report RR387 outlines damage scenarios that need to be considered in the assessment of damage stability with due consideration into potential collision scenarios as well as accidental flooding due to loss of watertight integrity. Loss of watertight integrity and resultant flooding can occur due to multiple reasons. The research program "RABL- Risk Assessment of Buoyancy Loss" undertaken during 1986-87 following the loss of Ocean Ranger and Alexander Kielland identified "ballast system failures and structural disintegration due to severe collision impact as the most dominant hazards which may cause buoyancy loss". Other causes include dropped objects, loosely secured mooring gear such as anchors and fairleads, poor integrity of ship side valves, unintentional operation of ballast valves due to control system failure or human error, material loss of watertight envelope or ship side attachments, failure of structural connections due to fatigue or corrosion, poor standards of watertight/weather tight integrity of openings etc.

Petroleum Safety Authority Norway (PSA) listed stability incidents for mobile offshore floating units since 2000 (Kvitrud, 2013 [7]). These include total loss of four units (P36, Aban Pearl, Jupiter 1, and Deepwater Horizon) between 2000 and 2010. Initiating events for these units vary, but they eventually sank following water ingress. A paper by PSA (Halsne et al., 2022 [8]) identifies a total of 13 major cracks, and 30 minor cracks related incidents in Norwegian Sector between 2000 and 2020 on semi-submersible units in addition to a high number of cracks rated as insignificant. These cracks have been reported among a fleet consisting of on average 18 mobile units and 6 to 8 production units. Major cracks are those that may threaten the integrity of the main load bearing structure. In addition, the paper also identifies five stability incidents of major severity resulting in severe listing. One of the reported incidents involved damages to multiple compartments (eight holes) by loosely secured anchor.

UK also has recorded multiple incidents resulting in loss of watertight integrity during the same period. Initiating events for semi-submersible units varied from fatigue cracks on braces and columns to damage from unsecured mooring attachments (fairleads) resulting in flooding of multiple compartments. For ship shaped units, water ingress was caused by cracks developed on shell plating and failure of ship side connections. Watertight subdivision of the hull, timely intervention by the crew and compliance with damage stability requirements prevented escalation under these scenarios. Table 1 provides a list of reported instances of loss of watertight integrity for floating offshore installations in UKCS since 2019.

Whilst there was at least one major collision between an offshore installation and attending vessel reported every year for the past three years (2021, 2022 and 2023) resulting in serious damage to the attending vessel, these did not involve a floating installation. For the past decade, there was a renewed focus on managing collision risk with attending vessels. Step change in Safety published guidance for Marine Operations: 500m safety zone [11] outlining good practice involved when vessels work within the 500m safety zone including safe approach and pre-entry checks, working alongside, and exiting the safety zone. The document acknowledges that having a 'marine responsible person' communicating with and monitoring the movements of the vessels within this 500m safety zone is good practice in the prevention of collisions. Even though it is difficult to conclude with such a small data set on recent collision incidents in UKCS, competent marine personnel onboard floating installations appear to have an influence in reducing collision with attending vessels. Other controls that prevent a collision risk include provision of detection systems such as Radar Early Warning Systems (REWS) and monitoring by Emergency Response and Rescue Vessels (ERRVs) deployed in the field. OEUK (OEUK, 2010, [19]) provides guidance on good practice and suggests benchmarks against which to assess compliance.

Table 1. Reported incidents of loss of watertight integrity in UKCS since 2019

Unit	Reported Year	Nature of Failure	Comment
FPSO	2019	Perforated overboard discharge pipe	Flooding of ballast tank
MODU	2019	Loss of buoyancy- two compartments	Impact from unsecured fairlead in heavy weather conditions
FPSO	2020	Multiple cracks on external hull plating and stiffeners	Diesel leak into sea
FPU	2020	Flooding of lift column	Poor integrity of non-return valves, operational errors
MODU	2021	Loss of buoyancy- single compartment	Impact from fairlead when mooring line tension lost following failure.
FPSO	2021	Two cracks appeared on shell plating	Diesel leak into sea
MODU	2021	Cracks on brace at connection to column	Water ingress of up to ~80 tonnes per day
MODU	2021	Cracks on a brace connection to column	Flooded brace and adjacent void space.
MODU	2022	Loss of watertight integrity- leak at weld on a brace	Flooded brace.
FPU	2022	One through thickness crack on column and six other cracks on similar locations.	Poor structural detail at brace connection to column, fatigue.
FSU	2022	Cracks on bottom shell plating	Through thickness resulting in seepage of sea-water into the tank

2.3 STRUCTURAL INTEGRITY

For a floating installation, maintaining structural integrity is closely linked to its ability to stay afloat, as demonstrated by the tragic incident involving Alexander Kielland in 1980 as well as numerous other incidents in the shipping industry. Incidents involving hull girder failure for ship shaped monohull offshore floating installations have been rare, unless initiated by an explosion in the cargo storage tanks (e.g. Trinity Spirit FPSO, 2022). However, for semi-submersible type units such as Alexander Kielland, hull structural configuration has played an important part in the total loss of the structure. Actual reasons for the loss of Aban Pearl, a mobile offshore drilling unit that was operating in Venezuela where it sank in 2010 are not well known, but there are images in the press of detached pencil columns floating after the incident indicating a structural failure either as initiating event or a consequence. One of the significant changes in design requirement brought by the Alexander Kielland incident is the requirement for structural redundancy i.e. the "ability to withstand the loss of any slender bracing member without causing overall collapse when exposed to environmental loading corresponding to a one-year return period for the intended area of operation" as laid out in the MODU code.

Structural design of a floater must take into consideration.

- Ability to withstand foreseeable loads including environmental loads and loading conditions,
- Ability to prevent progressive failure following accidental events such as ship collision, wave impact etc.
- Layout and configuration providing adequate redundancy against failure of a primary girder or slender bracing member.
- Adequate margin against fatigue failure to account for cyclic loading of primary structural members and inspection methodology (scope, frequency etc.) and
- Temporary phases during construction, transportation, installation, and decommissioning.

Figure 1: Pencil Column of Aban Pearl (Courtesy: https://diario16plus.com/)

Classification society rules have evolved to provide suitable guidance on global strength design and assessment for most of the standard configuration of floaters. However, there are often inconsistencies between these rules for certain aspects such as wave impact loading (Gopalakrishnakurup, 2022 [14]) that may affect the integrity of emergency equipment and escape routes. Such site specific or installation specific hazards may require a robust risk assessment and consideration of hierarchy of risk control measures in order of preference such as inherently safe design, prevention, control, and mitigation of hazards. Any novel design or configuration will also require evaluation of failure modes and risk reduction measures at design stage, eliminating or reducing hazards at source, through design.

Hull structural integrity issues almost inevitably lead to loss of watertight integrity of a floating installation, and hence affect stability as evident from some of the failures reported in Table 1.

2.4 POSITION KEEPING

For most permanent floating offshore installations, position keeping is provided by mooring lines constituting combinations of chain, wire, fibre ropes and other assemblies, often assisted by thrusters for heading control or load reduction. Survivability of the system under extreme environmental loads are assured by designing against the survival conditions with a suitable factor of safety stipulated in design codes. API RP 2SK, ISO 19901-7 or DNV-OS-E301 are all widely accepted design standards. All these three standards require a design check with one-line failed condition, thus ensuring adequate redundancy in the system, and preventing a cascading failure. Even though it is not a requirement under above mentioned design standards, checks for conditions above ULS environment have become commonplace for permanent installation to ascertain robustness of the system.

Whilst the target annual probability of failure of a single mooring line under DNV-OS-E301 is 10^{-4}, actual failure statistics would indicate much higher frequency of mooring failure, both among permanent and mobile offshore units. A study of mooring failure statistics (Brindley et al, 2014 [15]) suggest that a typical mobile drilling unit operating in UKCS experience a mooring failure in heavy weather every 24 operating years. In the period between 2000 and 2020, PSA recorded one triple line failure, one double line failure and 28 single line failure during MODU operations in Norwegian Continental Shelf (Halsne, et al, 2022 [8]). In UKCS, two double line failures were observed in 2021 and 2022, in addition to other single line failures. One of the double line failures occurred due to over-trawling of mooring lines outside 500m safety zone. Loss of tension on mooring lines has on two occasions resulted in damage to hull leading to water ingress into multiple compartments due to freely rotating fairleads impacting the hull. Mobile units operating in UKCS have also witnessed failure of winch gypsy wheel, displaced fairlead, failure of winch brakes among other common mode of failures associated with mooring chains. List of reported mooring failures in UKCS since 2019 is given in Table 2 below. These failures have been reported from a small population of mobile offshore units operational in UKCS. A new mode of failure with brittle fractures has also started emerging on R5 grade rental chain both in UK and Norway.

An industry survey of past failures on permanent installations (Fontaine et al., 2014 [17]) concluded that the annual failure rates for a single line, based on survey results, are an order of magnitude higher than typically assumed in mooring codes, arising from the design, construction, manufacturing, and installation phases. The study also concluded that overall rates of multiple line failure were found to be around 3.5 x 10^3 per facility compared to target failure rate of 10^{-5} [16]. Failure data gathered indicates that a multitude of causes from manufacturing to installation affect the integrity of mooring lines, and the design must cater for failure of mooring lines during operation to prevent failure of station keeping, and integrity management must be considered during the design process itself. A significant proportion (22%) of the failure inception occurred during installation.

Quality control during manufacturing is an important barrier that prevents mooring line failure and hence loss of position. Latent defects can result in catastrophic failures years after the installation is in service as evident in one of the recent incident in UKCS (IADC-Safety-Alert_6_22) when a chain lifter was split into two halves during operation. Lack of fusion in the flash butt weld of mooring chain links and inadequate heat treatment are common causes of mooring chain failures during service.

Table 2. Reported mooring failures in UKCS since 2019

Unit	Reported Year	Failure Type	Failed component	Comment
MODU 1	2019	Uncontrolled Release of Chain	Winch Brake	Brake band incorrectly set
MODU 2	2020	Loss of lower fairlead	Fairlead securing pin	Damage to fairlead securing pin
MODU 3	2021	Single Line failure	Chain link	Corrosion assisted fatigue
MODU 3	2021	Single Line failure	Chain link	Corrosion assisted fatigue
MODU 1	2021	Double Line Failure	Wire Rope	Did not prepare for storm conditions- High stress cycles during storm
MODU 4	2022	Winch Chain lifter split in two halves	Winch chain lifter	Latent defects from manufacturing
MODU 1	2022	Double Line Failure	Fibre Rope	Over-trawling
MODU 5	2022	Single Line	Connector (Kenter)	Kenter Link Failure at buoy attachment under low tensions during skidding
MODU 4	2023	Single Line failure	Chain link	Under low tension during mooring operation, failure mechanism to be established.
MODU 6	2023	3 Mooring lines parted during anchor proof-tensioning	To be confirmed	R5 grade chain anomalies (Pending Confirmation)

Even though mooring failures continue to be reported on mobile units, no mooring line failure has been reported in UKCS for a permanent installation since April 2017. Operators have managed to identify and remedy anomalies that threaten the integrity of mooring system before a failure has manifested on the system. Mooring integrity risk assessments in accordance with OEUK Mooring Integrity Guidelines (OEUK, 2014 [18]) have helped duty holders to develop an installation specific inspection program for the mooring system and implementation of a mooring integrity management system. In addition to a robust inspection program, several mitigating measures are also implemented for early detection of a loss of one or two mooring lines or to prevent loss of position in the event of multiple failures. These include provision of monitoring systems for position, line tension monitoring or line tension failure detection arrangements, and provisions for emergency towing arrangements in case of loss of multiple mooring lines.

Figure 2: Failure of a lower fairlead from latent defects from manufacturing.

3. MAINTAINING INTEGRITY

An important provision of DCR is the requirement for duty holders to ensure suitable arrangements in place for maintaining the integrity of the installation throughout its life cycle. These include arrangements for periodic integrity assessments and carrying out remedial work in the event of damage or deterioration that may prejudice its integrity. Such arrangements include planned maintenance and inspection of structures; periodic assessment of an installation taking account of its condition in relation to the original design expectations, assessment of damage or suspected damage, and arrangements for repair work in the event of damage or deterioration. A poorly maintained component can seriously affect the integrity of the installation. Figure 3 shows an example of poorly maintained mooring winch. Lack of inspection and maintenance of ship side valves and pipes can lead to water ingress as demonstrated from the incidents listed in Table 1 and Table 2.

Maintaining maritime integrity of a floating offshore structure often involves multiple inspection and maintenance activities often managed by different disciplines or even different entities specialised in such scope of work. These include:

- Routine preventive maintenance of onboard equipment such as pumps, valves, winches etc
- Calibration and testing of marine instrumentation.
- Inspection of internal tanks and void spaces
- Inspection of external hull above water line
- Inspection of underwater hull envelope including braces and joints as applicable and
- Inspection of mooring lines, connectors, and attachments (eg. subsurface buoys)

It is important that the inspection records are reviewed by competent personnel who have a good understanding of the failure modes and integrity issues associated with the system and its components based on the inspection findings and operational history. Quite often a new failure mode can manifest within the system or elsewhere, hence it is important that periodic assessments take in to account current good practices and incorporate advances in knowledge and changes in risk levels.

Most duty holders of bottom fixed structures have well established Structural Integrity Management (SIM) arrangements.

However, for floating structures, a mix of Class inspection and risk-based inspection arrangements are generally adopted. Scope and frequency under a prescriptive Class inspection regime developed for ships that are regularly drydocked do not always address integrity threats specific to the installation or location and can fall short of the regulatory expectations especially from a Major Accident Hazard perspective. Establishment of an integrity management framework in line with industry standards such as ISO 19901-9 [20] will be necessary to identify, detect and remedy the onset of integrity threats associated with the installation. This process as shown in Figure 4 is reflected in HSE offshore information sheet 6/2019 – Structural integrity management (HSE, 2019 [10]) HSE Inspection Guide on Maintenance Management (Maintenance

Management Inspection Guide) provides an overview of regulatory expectations on maintenance management arrangements.

Figure 3: Poor Maintenance and set up winch brakes.

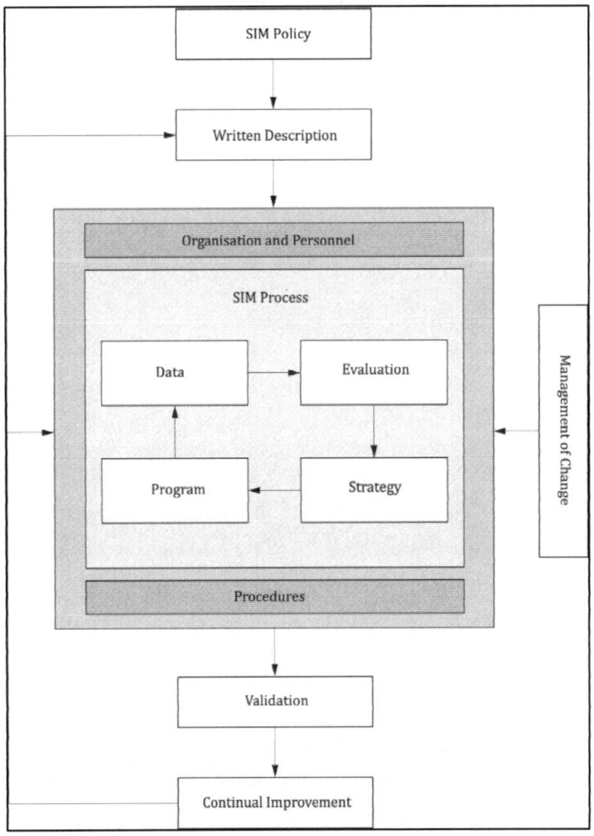

Figure 4: Integrity Management Framework (ISO 19901-9).

7. CONCLUSIONS

A floating structure must be designed for the following global failure modes, be able to withstand foreseeable loads on it and foreseeable damage.

a) capsizing
b) sinking
c) structural failure and
d) position keeping failure.

Reported incidents/accidents in UK & Norway indicate that failure of structural and mooring components as well as accidents resulting in loss of watertight integrity (eg. collisions, dropped objects etc) occur at a higher frequency than envisaged in design codes. So, the design of these structures must cater to the requirement that any such single event does not lead to a cascading catastrophic disaster. These principles would be applicable for any floating offshore structures including FOWT.

It is therefore emphasised that redundancy in structural and mooring configuration as well as ability to sustain damage to buoyant hull are important considerations that should form part of the design basis of a floating structure. A drifting floating structure is risk to not just people on it, but also to people on other installations, surface and subsurface infrastructure, ships, environment, and energy security.

8. ACKNOWLEDGEMENTS

I would like to thank my colleagues at HSE, especially Keith Pemberton & Satheesh Manayankath for reading through the paper and offering valuable comments.

The content of this paper reflects the opinion of the author, and not necessarily the official position of Health & Safety Executive, UK.

9. REFERENCES

1. A guide to the integrity, workplace environment and miscellaneous aspects of the Offshore Installations and Wells (Design and Construction, etc) Regulations 1996

2. A Guide to the Offshore Installations and Pipelines Works (Management and Administration) Regulations 1995, Guidance on Regulations

3. MOAN, T., 'Safety of Offshore Structures', *Core Report No.2005-04*

4. HSE Research Report RR387, "Stability", 2005

5. HSE Research Report RR143, "Review of the risk assessment of buoyancy loss (RABL) project", 2003

6. HSE Research Report RR473, "Review of issues associated with the stability of semi-submersibles", 2006

7. KVITRUD, A., "Modifications of the PSA Regulations Based on Case Studies of Stability Accidents", *OMAE2013-10041*, 2013

8. Halsne, M., Oma, N., Ersdal, G., Langoy, M., Andersen, T., Bjorheim, L.G., "Semisubmersible in Service Experiences on the Norwegian Continental Shelf", OMAE2022-81289, 2022

9. HSE Offshore Information Sheet No. 4/2011, "Flooding risk to machinery spaces of floating offshore installations: Guidelines on inspection of ship side valves; flood detection and control; inspection and training", 2011

10. HSE Offshore Information Sheet No. 6/2019, "Structural Integrity Management", 2019

11. Step Change in Safety, Marine Operations: 500m safety zone (Step Change in Safety | Marine Operations: 500m safety zone).

12. Petroleum Safety Authority, Norway, "Report from the investigation of the listing of Floatel Superior, 7 November 2012 on the Njord field", 2012

13. Code for the Construction and Equipment of Mobile Offshore Drilling Units (MODU Code), International Maritime Organization (IMO), 2009

14. GOPALAKRISHNAKURUP, R., "Wave Impact on semi-submersible units: a review of design standards", *OMAE2022-81456*, 2022

15. BRINDLEY, W., COMLEY, A.P., "North Sea Mooring Systems: How Reliable Are They?", *OMAE2014-23395*, 2014

16. Okkenhaug, S., Horte, T., Paulshus, O., "Summary and Recommendations for Safe Mooring System Design in ULS and ALS", OMAE2017-61534, 2017

17. FONTAINE, E., KILNER, A., CARRA, C., WASHINGTON, D., M A, K T., PHADKE, A., LASKOWSKI, D., KUSINSKI, G., "Industry Survey of Past Failures, Pre-emptive Replacements and Reported Degradations for Mooring Systems of Floating Production Units", *OTC-25273-MS*, 2014

18. OEUK, "Mooring Integrity Guidelines", Issue 3, 2014.

19. OEUK, "Guidelines for Ship/Installation Collision Avoidance", Issue 2, 2010

20. ISO 19901-9:2019, Petroleum and natural gas industries — Specific requirements for offshore structures — Part 9: Structural integrity management.

9. AUTHORS BIOGRAPHY

Renjeev Gopalakrishnakurup holds the current position of Specialist Inspector at UK HSE. The Author is responsible for regulating maritime integrity of floating offshore installations in UKCS and a nominated UK expert for floating structures and station keeping at the BSI Committee for the Design of Offshore Structures.